Book B1

THINKING CONNECTIONS

Concept Maps for Life Science

Frederick Burggraf

© 1998
CRITICAL THINKING BOOKS & SOFTWARE
P.O. Box 448 • Pacific Grove • CA 93950-0448
www.criticalthinking.com • Phone 800-458-4849 • FAX 831-393-3277
ISBN 0-89455-701-7
Printed in the United States of America
Reproduction rights granted for single-classroom use only.

During the writing of this book, I found myself indebted to a number of individuals for their wonderful help and support. I owe enormous thanks to Cheryl Block of Critical Thinking Books and Software for her constant devotion to the project, her many great ideas and her valiant suffering of my many opinions and notions. Also thanks to Carrie Beckwith at Critical Thinking for precise testing and retesting of the materials. I appreciate the work and content expertise of Dennis Holley and Mark Hull, classroom teachers who reviewed the maps carefully for factual and conceptual integrity and who had great suggestions. Thanks also to Dave May, Jim Kurtz and Dan Preotescu for their unfailing support throughout the project. And finally, adoring thanks to my wife, Karen, for being both cheerleader and champion of the project.

*I dedicate this book to the memory of my father,
Robert W. Burggraf,
a gentle man and educator who inspired me to be a teacher and
helped me to learn what pride there is in working with the
minds of the young.*

*Frederick Burggraf
February, 1998*

TABLE OF CONTENTS

CONTENT	TOPICS	MAPS		CONCEPT FILES	ANSWER KEYS
PRACTICE	Practice Map	2			3
		Lower Challenge	Higher Challenge		
CELL BIOLOGY	Cell Structures	4	5	6	7
	Cell Energy	8	9	10	11
	Cell Reproduction	12	13	14	15
	Levels of Organization	16	17	18	19
	Bacteria	20	21	22	23
	Viruses	24	25	26	27
	Protista	28	29	30	31
	Sarcodines	32	33	34	35
	Ciliates	36	37	38	39
	Flagellates	40	41	42	43
	Sporozoans	44	45	46	47
PLANT BIOLOGY	Algae	48	49	50	51
	Fungi	52	53	54	55
	Mosses	56	57	58	59
	Vascular Plants	60	61	62	63
	Ferns	64	65	66	67
	Roots	68	69	70	71
	Stems	72	73	74	75
	Leaves	76	77	78	79
	Flowers	80	81	82	83
ANIMAL BIOLOGY	Sponges	84	85	86	87
	Coelenterates	88	89	90	91
	Worms	92	93	94	95
	Mollusks	96	97	98	99
	Arthropods	100	101	102	103
	Cold-blooded Vertebrates	104	105	106	107
	Warm-blooded Vertebrates	108	109	110	111
HUMAN BIOLOGY	Skeletal System	112	113	114	115
	Muscular System	116	117	118	119
	Digestive System	120	121	122	123
	Circulatory System	124	125	126	127
	Respiratory System	128	129	130	131
	Excretory System	132	133	134	135
	Nervous System	136	137	138	139
	Human Reproduction	140	141	142	143

INTRODUCTION

Background

Concept maps were first developed from the cognitive learning theory of David Ausubel in the 1960s and 1970s and later popularized by Joseph Novak and D. Bob Gowin in their book *Learning How to Learn*. Fundamental components of this work include the following ideas:

- Knowledge is constructed.
- The most important factor influencing learning is what the learner already knows.
- Concepts are the central elements in the structure of knowledge and the construction of meaning.
- Concept maps serve to externalize concepts and improve thinking.

As graphic organizers, concept maps show how concepts (things, ideas, objects, activities) are related or linked to each other. They are visual road maps of meaning and understanding.

Concept maps model how people think: in hierarchies, in propositions, in contexts, and with multiple cross connections. The human mind is amazingly adept in seeing relationships in unexpected and wondrous ways, and can slip in and out of contexts and hierarchies with astonishing ease. Mapping of concepts allows the "mapper" to represent his or her understanding accurately, including far-flung links. Out of all of this, the mapper will construct and clarify conceptual *meanings*.

Thus, the best concept maps are those that students generate by themselves. Each student has a slightly (or not so slightly) different view and grasp of the concepts, and those differences will show up on the maps students create. Student-generated maps are, therefore, unique and intensely personal, and have tremendous value to both the students (whose cognitions are represented) and to you (it gives you a direct look into each student's understanding).

Unfortunately, bringing students to the point where they can do their own accurate, readable maps requires time, instruction, and lots of practice (rare commodities in today's classrooms). The next best alternative is to use concept map exercises such as those found in this book. These exercises have solid educational value in concept clarification, content reflection, meaning building, and assessment and foster many of the same skills required to originate concept maps.

About These Exercises

The concept maps in Thinking Connections span 35 topics in middle school life science. The topics were chosen using the National Science Education Standards as a guide.

The practice exercise on page ix is a good introduction to concept mapping. It uses terms and concepts that should already be familiar to students. This exercise will give students the idea of what the tasks ahead will be in content-rich maps.

Concept maps can be used with a wide range of student abilities because they represent the individual understanding of the learner, regardless of his or her level. There are two levels for each map topic, addressing the needs of a broader population of students. The lower challenge map is designated by *two* small bars at the end of the title line; the higher-challenge map has *three* small bars at the end of the line.

Typically, lower-challenge maps have:
- more seed words,
- fewer technical words,
- shorter word lists,
- fewer connectors and simpler paths.

The presence of two levels of maps allows you to differentiate your assignment, even in the same class. It also allows these maps to be used across a wide spectrum of abilities and grade levels.

Structure, Features, and Notations of Concept Maps

Concept maps have three major components: items, connectors, and labels.

❐ *Items*—These are objects, ideas, places events, processes, and activities. Item names are placed in boxes in the concept map. The "main" item of a map is always shown in all capital letters. It sets the *domain*, or topic, of the map. Other items are lower case, showing subordinate relationships. Concept maps typically follow a top-down hierarchy of organization, with the main item being a general term and descending terms becoming increasingly specific. Students read maps intuitively this way, starting at the top and working downward. (Unfortunately, the limits of forcing a map onto one sheet of paper often compromise aspects of the hierarchy.)

- ❏ *Connectors*—Connectors are lines that join items, indicating which ones are related or linked and how they are linked. Connectors provide map syntax. The maps in the book were designed to be read from top to bottom (and, to a lesser degree, from left to right). The natural top-down flow eliminates the need for every connector to be an arrow.

- ❏ *Labels*—Labels are words associated with each connector. They show *how* items are related or linked.

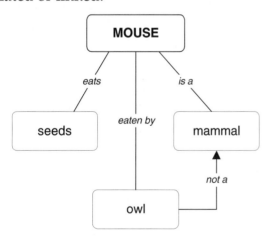

Figure 1. A simple concept map

The concept map in Figure 1 has four items. The lines that connect them are connectors. Each connector has a label showing the nature of the link.

Arrows

The natural flow of the concept maps in the book is top-down, left-to-right. In the map below, for example, it's obvious that the representation is that the vulture eats the raccoon.

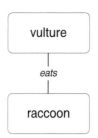

Figure 2. Top–down interpretation is intuitive

There are instances, however, where there is ambiguity about the direction of the relationship. For example, in Figure 3, do vultures eat (dead) raccoons or do raccoons eat vultures (eggs)?

Figure 3. In some cases, interpreting relationships is difficult

Arrows show how to read the relationships and links. In Figure 4, the arrow makes it clear that, in this case, raccoons eat vultures.

Figure 4. Adding arrows simplifies interpretation

The Practice map (pg. xi) has three arrows; when you work that map with your class, discuss how the arrows show the direction of the relationships.

Line Styles

The normal style of a connector is a solid line. A dotted line may be used to show a weaker, more tentative or less likely link.

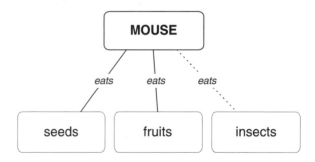

Figure 5. A dotted line shows a weaker link

For example, in Figure 5, a mouse is normally an herbivore but, in some rarer situations, may eat other animals. The dotted line shows this relationship.

A dotted line may be used to clarify relationships. In Figure 6, the connector *to* doesn't actually define the relationship, but rather clarifies the connector *attach*.

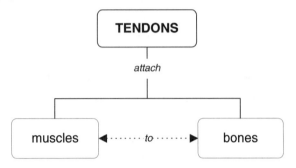

Figure 6. The dotted connector helps explain another connector

Multiple Connectors

Connectors can be joined to create multiple connectors. These usually take two forms:

1. The "fork" connector, branching out forklike under an item.

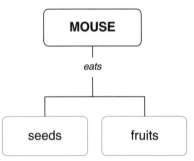

Figure 7. A "fork" connector

Notice how, in this case, the label *eats* appears only once and defines both relationships.

2. The "flag" connector, branching out on one side under an item.

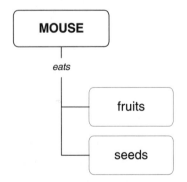

Figure 8. A "flag" connector

In both examples, the relationships are read as "and" relationships: "a mouse eats fruits *and* seeds."

There are times, however, when relationships are exclusive and the word *or* would be correct. For example, a warm-blooded animal has either feathers or fur, but not both. To show this in a map, another notation is needed:

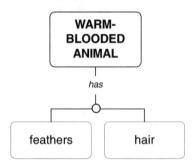

Figure 9. The circle symbol represents "or"

The small circle added here changes the interpretation to "or." One use of this symbol in this book appears on the Sponges map (pg. 82).

Seed Items

All the maps in this book include a few "seed" items—words already placed in their correct locations on the map. Lower-challenge maps typically have more seed words than higher-challenge maps.

The seed words, in combination with surrounding connector labels, provide clues to get students started filling in the map. The map answer keys suggest good starting points.

Colors

A number of maps in this collection ask students to color certain items and then complete a Color Legend. This allows for another layer of information to be represented on the maps. Students will need highlighters, colored pencils, or crayons. Correcting mistakes made in coloring maps is difficult, so have spare maps handy.

Actually, any map can be made better with color, even if it's just to color in various branches to make them more distinguishable. Any time a student colors branches on a map, however, encourage him or her to make a Color Legend so that the meaning, no matter how simple, is clear.

Shapes

A few maps in this collection use the shape of the item boxes to represent layers of information (the Levels of Organization map on pg. 14, for example). In those cases, students are required to follow a Shape Legend.

Picture Galleries

At their very core, concept maps are graphic organizers, and the presence of drawings or other visual representation of concepts adds to their abilities to represent learning. Some of the maps in the book have collections of pictures for the student to place in the map.

Activities that involve graphics may require scissors and tape or glue in the classroom. Some students may balk at cutting out the graphics and taping or pasting them on the map, but later on, when it's time to use the map to review the topic, those graphics will make the map even more understandable and accessible.

Diagrams

Some of the maps in this book ask students to match parts of a diagram to individual items. Certain items have a box in one corner for the student to write a letter corresponding to a letter on the diagram.

The Concept Files

Concept Files are review summaries included in this book for each topic. The Concept Files cover topic background, vocabulary, major informational points, and other important data. Covering all the map content, Concept Files allow you to use all the maps in this book regardless of the curriculum you're using.

Using the Concept Files

If the curriculum you're using does not include all the concepts and vocabulary found on the map, give students the Concept File for the topic and let them use it as a study guide. Alternatively, you could provide other references and resources (textbooks, dictionaries, encyclopedias, etc.) so that students can look up the terms not in your course text.

The Concept File may make a good review sheet for a topic even if you don't use the concept map.

How to Use the Concept Map Exercises

Concept Maps are extremely versatile across a wide range of student abilities. Here are just a few suggestions for using these concept maps:

1. **As study aids**—Students who ask for extra help may benefit if they are given a map to complete while they read the source materials.

2. **As assessments**—Concept maps are excellent assessment tools and can be used to evaluate student understanding as well as the effectiveness of your teaching. Students who have a firm conceptual grasp of the material will usually do very well with the exercises. When a student has a lot of difficulty with a map, it usually indicates a lack of understanding of the material. (Oddly enough, students who do very well at memorizing may have trouble with concept maps. This is because these students have learned to do well by echoing back material without really understanding it.) If many students consistently show errors in one or more parts of a map, it may be a sign to go over the material again or in a different way.

3. **As review**—Before the end of a unit, use the concept map as a class activity for reviewing the lesson. Also, previously completed maps make great starting points as students study for large unit, semester, and course tests.

4. **As homework**—As students read new materials or study for review, concept maps can provide a focus activity and help the students graphically organize the content.

5. **As small-group activities**—Allowing students to work in groups of two or three can produce interesting and valuable interactions. You can allow students to use their textbooks and other references, or make the activity a closed-book exercise. (If you do allow references during map activities, you'll find that many groups will use these resources.) Typically, students will plan together and consider various possibilities for completing the map. It's very common to hear students arguing points of concept placement and the logic behind them. Stu-

dents also get intense practice with vocabulary during these sessions. Most importantly, the *meaning* and *understanding* of concept becomes paramount; it isn't that they just *use* the words and ideas, but rather the map helps them *construct* meaning and understanding.

One strategy that can get students intensely involved is to correct a group's map right away and say, "There are [some number] items incorrect, but I'm not going to tell you which ones. Keep working." The students will go back and attack the map with new vigor, new arguments, and new thinking.

6. **As portfolio materials**—If you're using portfolio assessment, the concept maps that students complete are great additions to the materials you or the student chooses to put in the portfolio.

7. **As parts of tests**—If your students have not used a map as another activity, you could attach it to a chapter or unit test.

8. **As writing exercises**—Once students complete their maps, ask them to write a paragraph using the map as a guide. Items in the map will usually become subjects and objects of the sentences; connector labels will be used as the verbs.

About the Author

Frederick Burggraf began his career as a high school science teacher, teaching two years in western Pennsylvania and ten years in southern Maryland. In the early 1980s, he left the classroom and devoted his full-time efforts to developing educational software. He has authored over 40 software titles published by three national software houses. Mr. Burggraf is currently a specialist in verbal and visual communications, as well as an educational consultant. He has a B.S. from Indiana University of Pennsylvania (1970) and an MEd from the University of Maryland (1983).

THE PRACTICE MAP

In the map exercises in this book, students will be asked to fill in missing items from a list. They may also be asked to supply connector labels, complete a map legend, match items with a diagram, or cut out and affix images on the map.

The first map in this book is a practice map and serves to introduce your students to concept mapping and several features of these exercises. Many students take readily to concept maps, intuitively knowing what to do. But others have trouble with them, especially at first. The practice map can clear up many questions if you take some time to work the map with the class.

Strategy

Here is one strategy for using the Practice Map:

1. Pass out the map and use an overhead transparency on a projector as you work through it.
2. Suggest that the students look over the entire map to see all its aspects and what is included. Have them read *all* the connectors and terms in the word list. Point out that there are also graphics to place in the map. *It's this "big picture" approach that some students miss and is vital to completing concept maps successfully.*
3. Have students cut out the graphics before they start filling in the map.
4. Ask students for suggestions on where to start. More importantly, *ask them how they came up with those suggestions.* In other words, find out what strategies they used to attack the map. If students need a jump start, ask them to notice the item *water*. Ask, "What on the list or in the pictures is related to water?" The picture of the *canoe* is the only candidate. Now ask, "What powers a canoe?" After students have filled in the item *person*, have them notice the arrow pointing to *person*, connecting back with another *powered by* connector. Ask, "What else is powered by a person?" Then, let them go for a while.

This process is typical in completing concept map exercises. Often, students begin on items near the bottom of the map (where the hierarchy is most specific) and then move up the map, checking against other connectors and item names. Although some items can be placed in several spots on the map, there's only one way that all the items fit together.

5. When the items are all placed, it's time to color the map. Students should use highlighters, crayons, or colored pencils—if you can't provide these, ask the students the day before to bring them in.

The topic of this map, "Means of Transportation," was chosen for its familiarity—students can concentrate on the dynamics of the map and not struggle with the conceptual content.

The answer key for the practice map has many other suggestions for its use.

THINKING CONNECTIONS: Life Science Book B   Practice Map

Concept Map: Practice!

Directions: Select words from the word list and fill in the blank map items. Use each word only once, and use all the words on the list. Cut and paste (or tape) the pictures in the correct boxes on the map. Then use highlighters, colored pencils, or crayons to color items that (1) have wheels and (2) are living. Show your color scheme in the legend.

Name _____
Date _____ Period _____

WORD LIST
- air
- bicycles
- carts
- engine
- land
- person
- trucks

COLOR LEGEND
- ☐ Has wheels
- ☐ Is living

PICTURE GALLERY (helicopter, canoe, car, hand/glove, motorcycle, space shuttle)

Central concept: **MEANS OF TRANSPORTATION**

Branches:
- some travel in the → [ ] → vehicle types include → [ ] cannot hover / [ ] can hover
- some travel on → water → vehicle types include → [ ] powered by [ ] ; example → [ ]
- some travel on → [ ] → vehicle types include → cars (example → [ ], runs on gasoline; have [ ]; have no [ ], powered by [ ]); have no [ ], pulled by [ ]

Answer Key

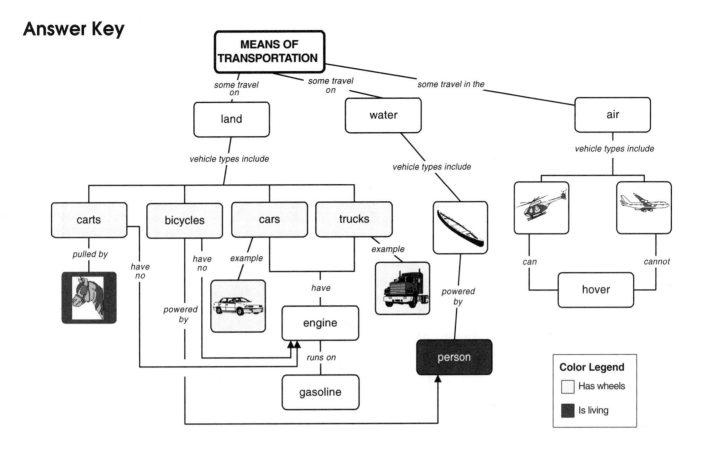

Teaching Suggestions

This practice map gives students experience working with a map where most of the information required is common knowledge. The map also introduces several of the map features found in this book.

Starting Hints

Students should first complete the map using the word list and the pictures. The item *water* is a good starting point because there's only one picture of a water vehicle (the canoe).

Some students may focus on the word list and forget that the pictures are choices for items in the map. In this map, square items hold pictures and rectangular items hold text.

Notes

Carts and the horses that pull them may be unfamiliar to some students. Including these items in the map was intentional; in many cases, students will have to finish maps by the process of deduction, finding a "best fit" for the items left over.

Once the maps are complete, you may want to check them for accuracy before you direct the students to color them. (Incorrectly colored items are hard to correct.)

Students should use one color to indicate those items that have wheels. These items are: carts, bicycles, cars, trucks, sedan (pictured), and semi (pictured). The picture of the jet was not included in this category because its wheels are not the primary method of movement. Nonetheless, it *does* have wheels, and some students may correctly argue the point. If they do, welcome it; this is an important part of the process of mapping—argument, defense, insight, and revision. That students are arguing a point and defending it is far more important as a critical thinking skill than getting the "correct answer."

Students should use a second color to indicate the items that are living. In this case, there are two: horse (pictured) and *person*.

A small number of students have serious difficulty with concept maps; this practice map should reveal who those students are and where they have problems.

Concept Map: Cell Structures

Directions: Select words from the word list and fill in the blank map items. Use each word only once, and use all the words on the list.

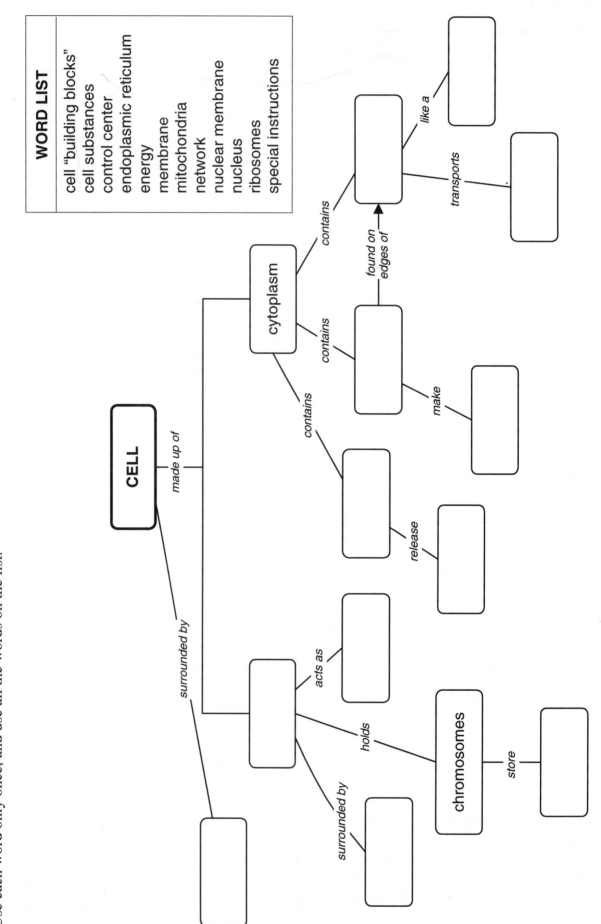

Concept Map: Cell Structures

Directions: Select words from the word list and fill in the blank map items. Use each word only once, and use all the words on the list. Then use two different highlighters, colored pencils, or crayons to color in items that are (1) strictly related to plants and (2) are related to both plants and animals. Show your color scheme in the legend.

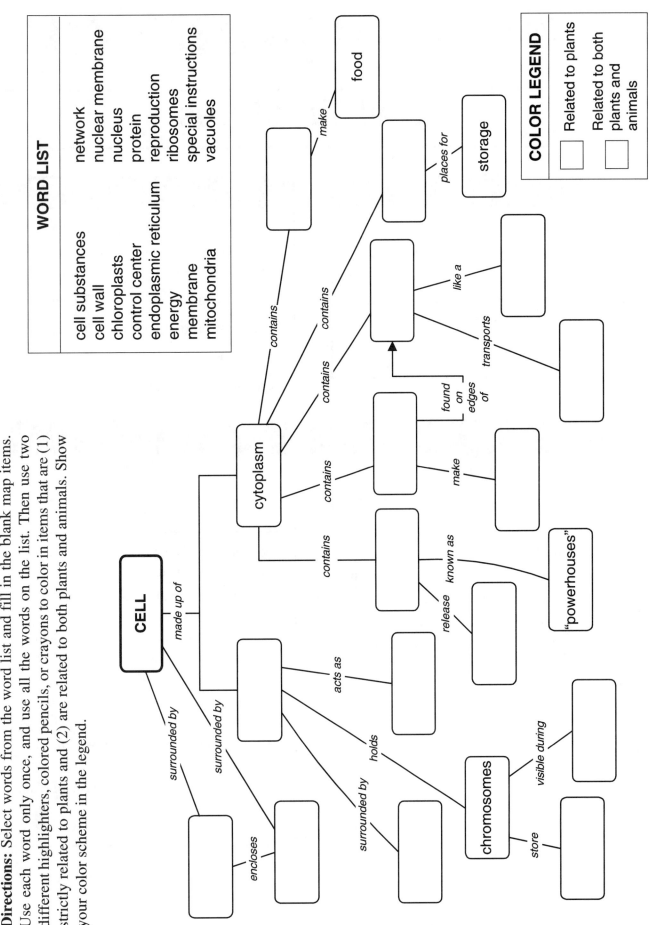

WORD LIST

cell substances	network
cell wall	nuclear membrane
chloroplasts	nucleus
control center	protein
endoplasmic reticulum	reproduction
energy	ribosomes
membrane	special instructions
mitochondria	vacuoles

COLOR LEGEND

☐ Related to plants
☐ Related to both plants and animals

THINKING CONNECTIONS: Life Science Book B — Cell Biology

Critical Thinking

Cell Structures

Background

Cells
- are the basic units of living things
- are protected by a membrane
- are also protected by a tough outer wall (plant cells)

Vocabulary

Check your understanding—these terms are explained on this page.

- ❐ **chromosomes**
- ❐ **nucleus**
- ❐ **cytoplasm**
- ❐ **protein**
- ❐ **endoplasmic reticulum**
- ❐ **ribosomes**
- ❐ **membrane**
- ❐ **vacuole**
- ❐ **nuclear membrane**

Two Main Areas of the Cell

Nucleus

The nucleus
- is surrounded by a nuclear membrane
- is the control center of the cell
- stores special cell information in chromosomes, tiny threadlike structures seen during cell reproduction

Cytoplasm

The cytoplasm is the place of many cell activities and structures.

- Mitochondria are the "powerhouses," storing energy.
- The "building blocks" of cells (proteins) are manufactured by ribosomes (located along the endoplasmic reticulum).
- Cell substances are transported on a network called the endoplasmic reticulum.
- Plant cells have vacuoles in which cell materials are stored.
- Food is made in the chloroplasts of plant cells.

THINKING CONNECTIONS: Life Science Book B — Cell Biology

Cell Structures

LOWER CHALLENGE

Score: 12 words

Starting hints: The connector *found on the edges of* is a good early clue. Also, note that something *holds* chromosomes.

Concept map (Lower Challenge):
- CELL
  - surrounded by → membrane
  - made up of → nucleus
    - surrounded by → nuclear membrane
    - acts as → control center
    - holds → chromosomes
      - store → special instructions
  - made up of → cytoplasm
    - contains → mitochondria
      - release → energy
    - contains → ribosomes
      - make → cell "building blocks"
      - found on edges of → endoplasmic reticulum
    - contains → endoplasmic reticulum
      - transports → cell substances
      - like a → network

HIGHER CHALLENGE

Score: 38 (16 words + 22 colored boxes)

Starting hints: The connector *found on the edges of* is a good early clue. Also, note that something *holds* chromosomes.

Notes: Plant-cell features are additions to this level.

Concept map (Higher Challenge):
- CELL
  - surrounded by → cell wall
    - encloses → membrane
  - surrounded by → (membrane)
  - made up of → nucleus
    - surrounded by → nuclear membrane
    - acts as → control center
    - holds → chromosomes
      - store → special instructions
      - visible during → reproduction
  - made up of → cytoplasm
    - contains → mitochondria
      - release → energy
      - known as → "powerhouses"
    - contains → ribosomes
      - make → protein
      - found on edges of → endoplasmic reticulum
    - contains → endoplasmic reticulum
      - transports → cell substances
      - like a → network
    - contains → chloroplasts
      - make → food
    - contains → vacuoles
      - places for → storage

Color Legend
- ■ Related to plants
- ▢ Related to both plants and animals

© 1998 Critical Thinking Books & Software • www.criticalthinking.com • (800) 458-4849

Concept Map: Cell Energy

THINKING CONNECTIONS: Life Science Book B — Cell Biology

Name _____
Date _____ Period _____

Directions: Select words from the word list to fill in the blank map items. Use each word only once, and use all the words on the list.

WORD LIST
chlorophyll
chloroplasts
life processes
mitochondria
photosynthesis
respiration
sugars

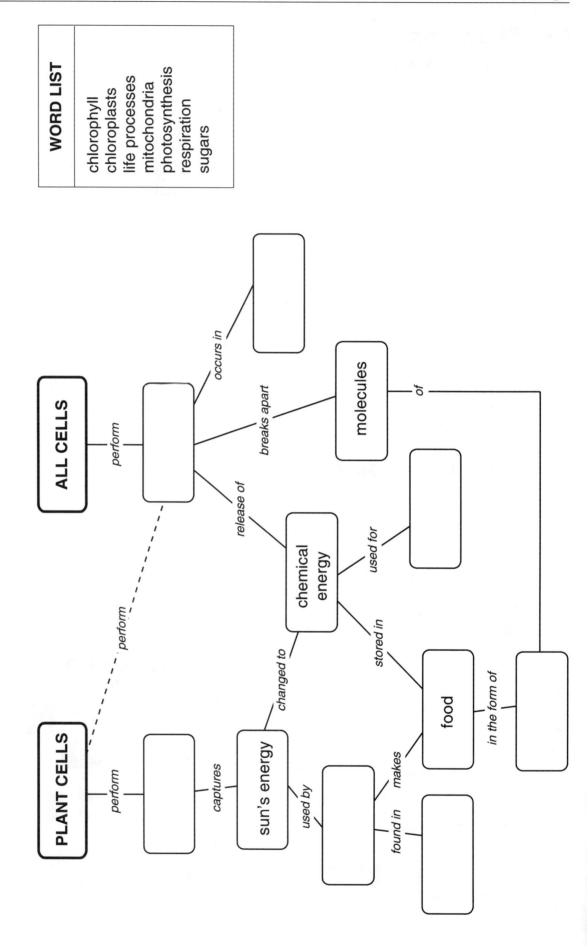

Concept Map: Cell Energy

Directions: Select words from the word list to fill in the blank map items. Use each word only once, and use all the words on the list.

WORD LIST

ALL CELLS
chlorophyll
chloroplasts
life processes
mitochondria
molecules
photosynthesis
PLANT CELLS
respiration
sugars
sun's energy

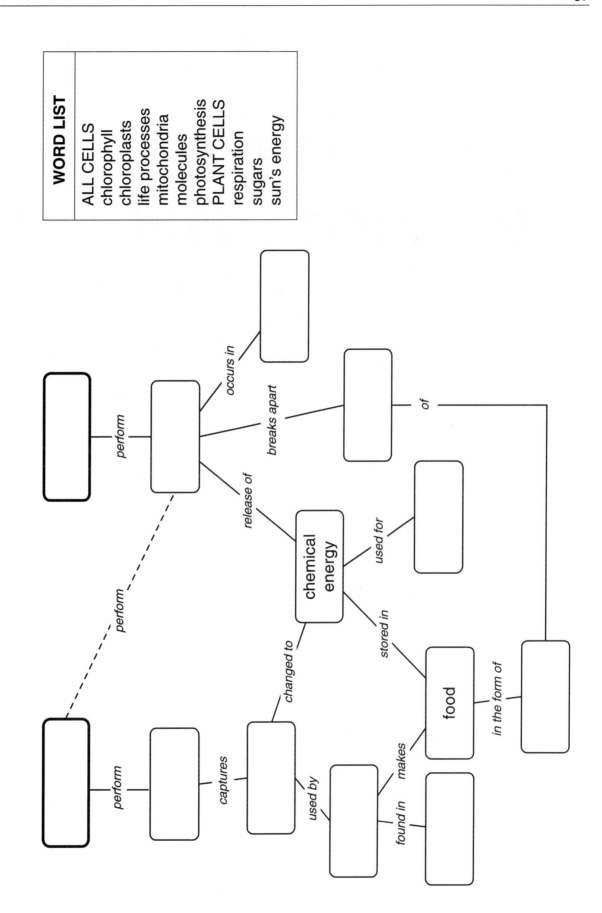

THINKING CONNECTIONS: Life Science Book B — Cell Biology

Critical Thinking

Cell Energy

Background

Cells require energy to live.

- Some cells (such as most plant cells) capture light energy and store it in foods such as sugar.
  - Chlorophyll is a substance that can capture light energy.
  - Chloroplasts in cells contain the green pigment chlorophyll.
- *All* living cells burn food and release energy.
  - In most cells, the burning of food occurs in the mitochondria.
  - Burning food releases its energy.

Vocabulary

- **chlorophyll**—A green pigment used during food manufacture.
- **chloroplasts**—The part of the cell where photosynthesis occurs.
- **mitochondria**—Centers of energy release in the cell.
- **photosynthesis**—The process by which living things make their own food.
- **respiration**—The release of energy during the burning of food.

Cell Energy
LOWER AND HIGHER CHALLENGE

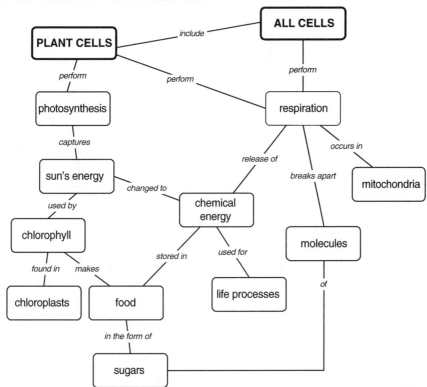

Score:
Lower Challenge 7 words
Higher Challenge 11 words

Starting hints: Begin at *food* and work up. On the higher challenge map, the main items are not filled in, but appear in the word list in capital letters. If students need help, suggest that the terms in all capitals will likely belong in the larger boxes at the top of the map.

Concept Map: Cell Reproduction

Directions: Select words from the word list to fill in the blank map items. Use each word only once, and use all the words on the list.

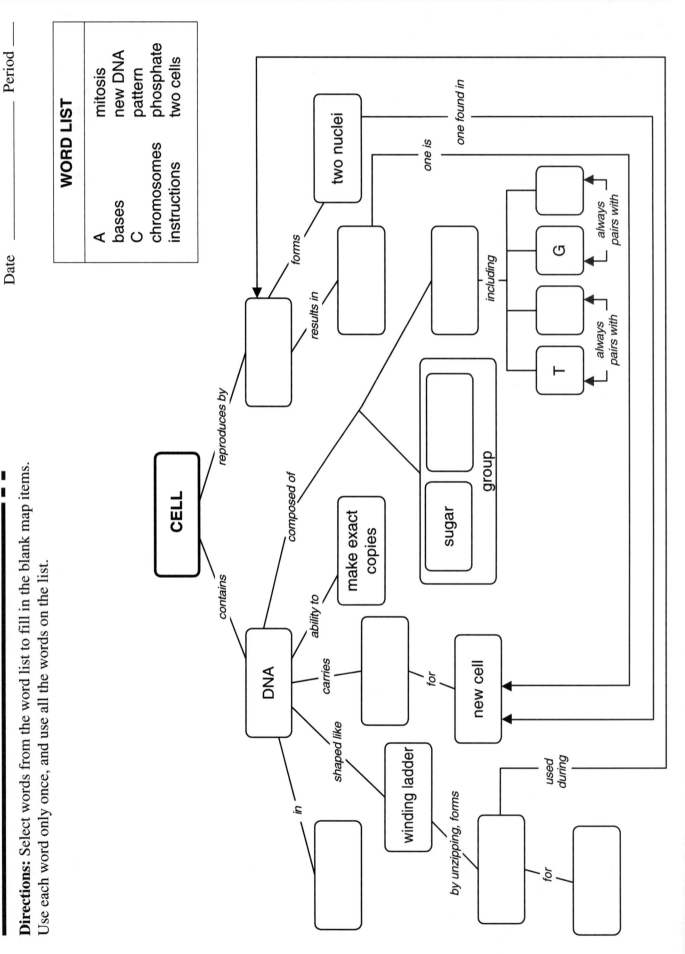

WORD LIST

A
bases
C
chromosomes
instructions
mitosis
new DNA
pattern
phosphate
two cells

Concept Map: Cell Reproduction

Directions: Select words from the word list to fill in the blank map items. Use each word only once, and use all the words on the list.

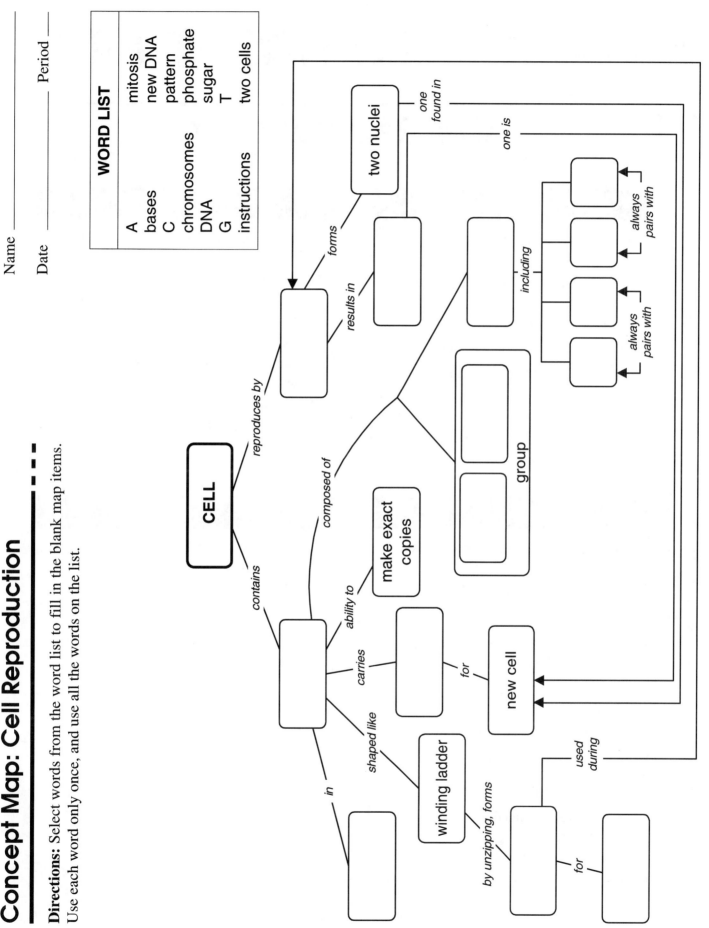

THINKING CONNECTIONS: Life Science Book B — **Cell Biology**

Critical Thinking

Cell Reproduction

Background

When a cell reproduces, it splits into identical copies of itself.

- The process of splitting is called *mitosis*.
- DNA guides the process.
- The cell nucleus divides in two.
- DNA creates a copy of itself.

Vocabulary

- ❒ **base**—One of four similar chemicals in DNA.
- ❒ **chromosomes**—Small threadlike objects in the cell's nucleus that contain DNA.
- ❒ **DNA**—A chemical that carries instructions for making a new cell; made of sugar-phosphate groups and bases.
- ❒ **mitosis**—The process where a cell divides into two cells, each with a complete nucleus.

DNA

DNA is the chemical deoxyribonucleic acid.

- It is shaped like a double winding ladder.
- When the ladder "unzips," it becomes a pattern to build two ladders.
- DNA is made of four nucleotide bases: A, T, C, and G (adenine, thymine, cytosine, and guanine)
  – A always pairs with T
  – C always pairs with G
- Each base is joined to sugar-phosphate group.

Mitosis

During mitosis, a cell divides and forms two identical cells.

- The cell's nucleus divides into two.
- The cell's DNA "unzips" and makes new DNA.

Cell Reproduction
LOWER AND HIGHER CHALLENGE

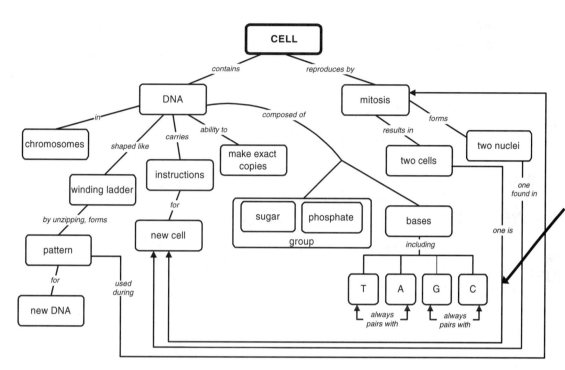

Score:
Lower Challenge 10 words
Higher Challenge 14 words

Starting hints: On the lower challenge map, the seed nucleotide bases T and G have obvious pairs.

On the higher challenge map, the seed nucleotide bases can be matched in pairs.

Notes: On the higher challenge map, the placement of *sugar* and *phosphate* can be reversed.

Also on the higher challenge map, the order of the pairs can be switched, as long as A is paired with T and C is paired with G.

Concept Map: Levels of Organization

Directions: Select words from the word list to fill in the blank map items. Use each word only once, and use all the words on the list. Place plant examples on the left (in the boxes shaped like ⏢) and animal examples on the right (in boxes shaped like ⬡).

SHAPE LEGEND
⏢ Plant feature
⬡ Animal feature

WORD LIST

bone	oak tree
brain	ORGANISMS
circulation	ORGANS
corn	ORGAN SYSTEMS
flower	root
kidney	squirrel
leaf	TISSUES
muscle	worm

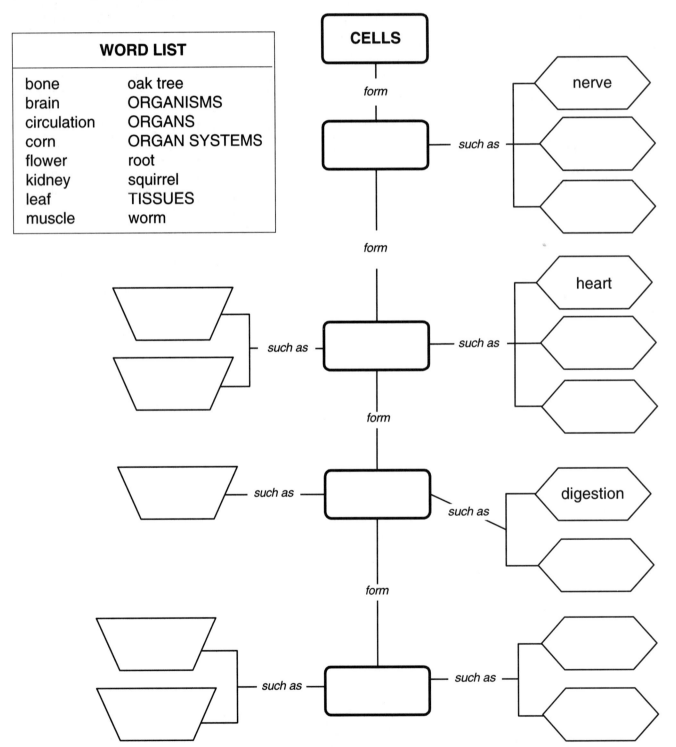

Concept Map: Levels of Organization

Directions: Select words from the word list to fill in the blank map items. Use each word only once, and use all the words on the list. Use the map legend to place plant and animal examples in the appropriate boxes.

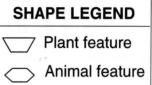

SHAPE LEGEND
- Plant feature
- Animal feature

WORD LIST

bone	muscle
brain	nerve
CELLS	oak tree
circulation	ORGANISMS
corn	ORGANS
digestion	ORGAN SYSTEMS
flower	root
heart	squirrel
kidney	TISSUES
leaf	worm

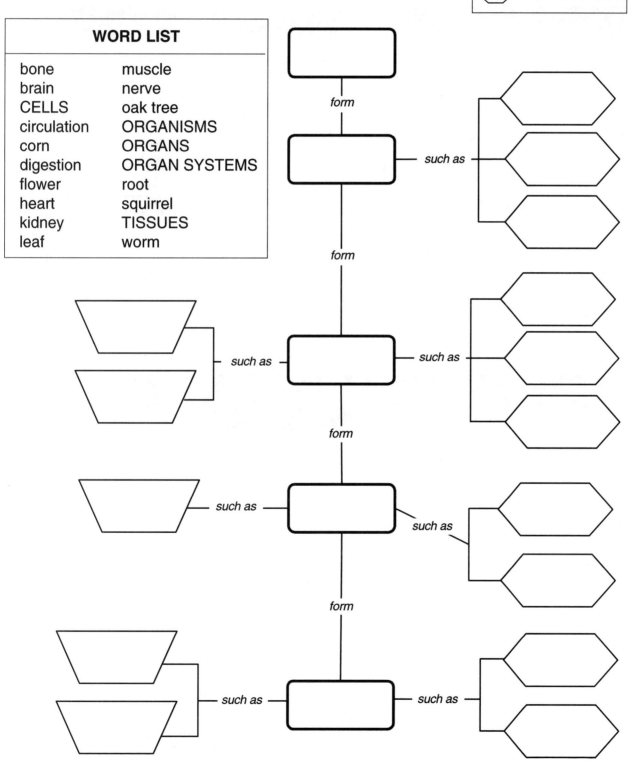

Critical Thinking → Concept File

Levels of Organization

Levels

The cell is the basic unit of life activity.

Cells that have similar functions and that work together make up tissues.

Tissues that have similar functions and that work together make up organs.

Organs that have similar functions and that work together make up organ systems.

Organ systems that work together make up organisms.

Vocabulary

- ❏ **organ**—A working group of tissues with similar functions.
- ❏ **organism**—An independent living thing; may be a single cell, often a working group of organ systems.
- ❏ **organ system**—A working group of organs with similar functions.
- ❏ **tissue**—A working group of cells with similar functions.

Tissues

Animal examples
- muscle
- nerve
- bone

Organs

Animal examples
- brain
- heart
- kidney

Plant examples
- leaf
- root

Organ Systems

Animal examples
- digestive system
- circulatory system

Plant example
- flower

Organisms

Animal examples
- squirrel
- worm

Plant examples
- corn
- oak tree

Levels of Organization
LOWER AND HIGHER CHALLENGE

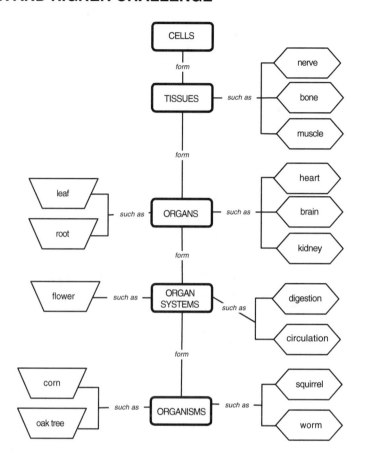

Score:
Lower Challenge 16 words
Higher Challenge 20 words

Starting hints: Have students place words that are all capitals first. These are main items that go in the larger boxes at the top and center of the map.

Suggest that students go through the list first and separate items according to plants and animals.

On the higher challenge map, suggest that students start at the bottom and work up.

Notes: When scoring, keep in mind that plant terms are on the left and animal terms are on the right. These items do not need to be listed in any specific order.

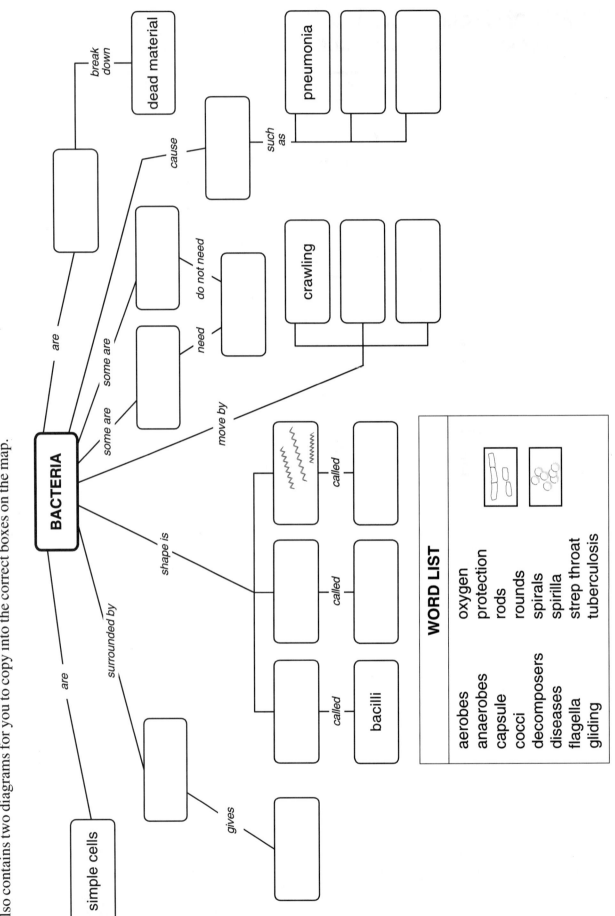

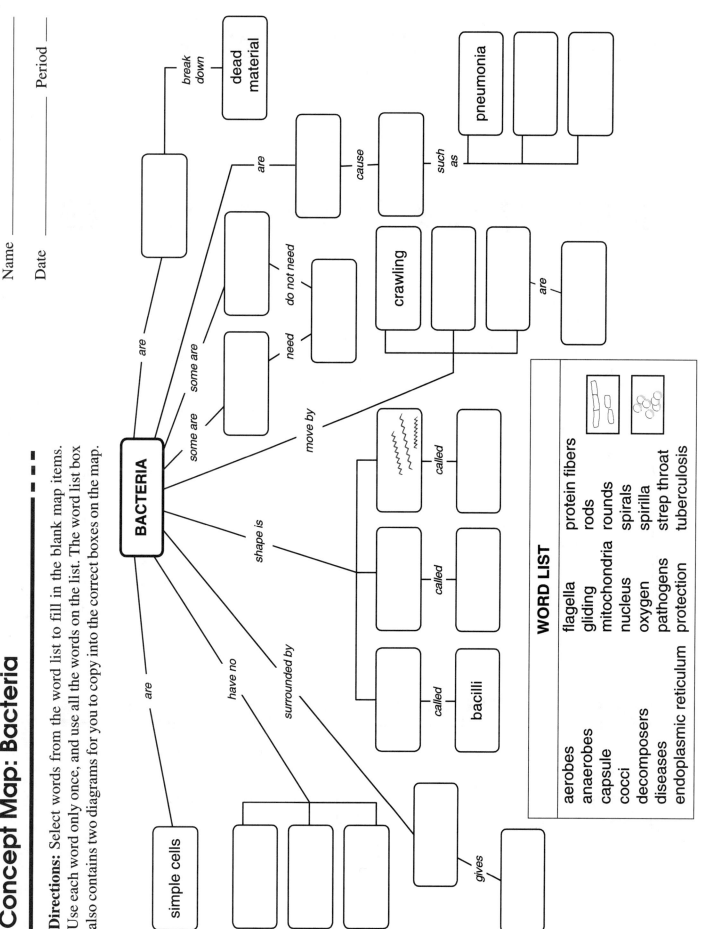

Critical Thinking

Bacteria

Background

Bacteria
- are simple cells without mitochondria, endoplasmic reticulum, or a nucleus
- are protected by a capsule on the outside of the cell

Some bacteria
- are decomposers
- are pathogens
  - strep throat
  - pneumonia
  - tuberculosis
- are aerobic
- are anaerobic

Vocabulary

- ❐ **aerobic**—Requires oxygen to live.
- ❐ **anaerobic**—Lives in the absence of oxygen.
- ❐ **bacillus** (plural bacilli)—A rod-shaped bacterium.
- ❐ **coccus** (plural cocci)—A round-shaped bacterium.
- ❐ **decomposer**—Feeds on dead material.
- ❐ **endoplasmic reticulum**—A connecting network in some cells.
- ❐ **flagellum** (plural flagella)—A whiplike structure used during movement and made of fibers of protein.
- ❐ **mitochondria**—Centers of energy release in some cells.
- ❐ **nucleus**—The control center of some cells.
- ❐ **pathogens**—Cause diseases.
- ❐ **spirillum** (plural spirilla)—A spiral-shaped bacterium.

Methods of Movement

Bacteria move in three ways:
- by crawling
- by gliding
- by using flagella

Types of Bacteria

Bacteria are classified by their shapes:
- round—called cocci
- spiral—called spirilla
- rods—called bacilli

THINKING CONNECTIONS: Life Science Book B — Cell Biology

Bacteria

LOWER CHALLENGE

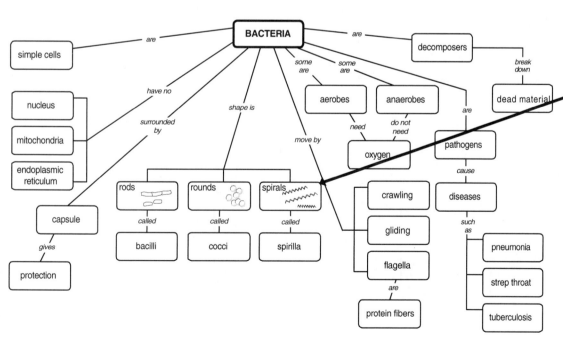

Score: 18 (16 words + 2 diagrams)

Starting hints: The seed item *pneumonia* suggests *diseases*. The seed item *bacilli* is the name given to *rods*.

Point out the small diagram of spirals on the map. Students should copy the other two diagrams from the word list, then label all three.

HIGHER CHALLENGE

Score: 23 (21 words + 2 diagrams)

Starting hints: The seed item *pneumonia* suggests *diseases*. The seed item *bacilli* is the name given to *rods*.

Point out the small diagram of spirals on the map. Students should copy the other two diagrams from the word list then label all three.

Notes: The word list includes five terms not used on the lower challenge map: pathogen, nucleus, mitochondria, endoplasmic reticulum, protein fibers.

THINKING CONNECTIONS: Life Science Book B — Cell Biology

Name _____
Date _____ Period _____

Concept Map: Viruses

Directions: Select words from the word list to fill in the blank map items. Use each word only once, and use all the words on the list.

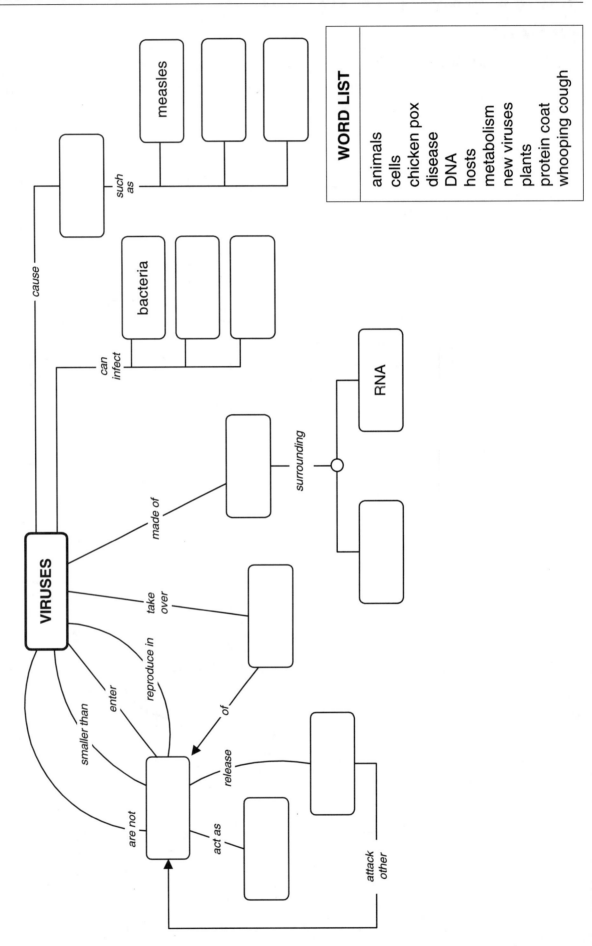

WORD LIST

animals
cells
chicken pox
disease
DNA
hosts
metabolism
new viruses
plants
protein coat
whooping cough

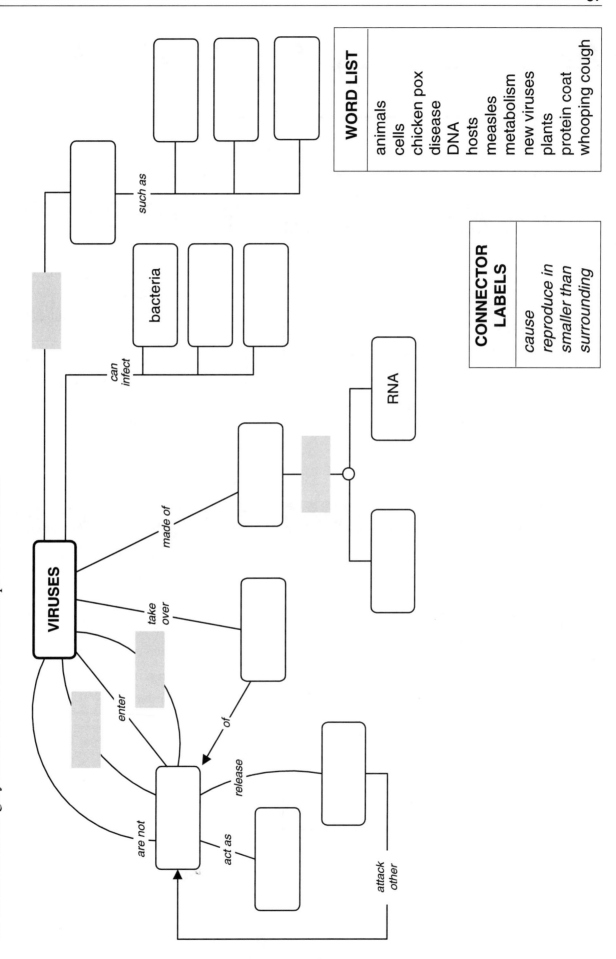

Critical Thinking

Viruses

Structure

Viruses have
- a core of either
  - DNA
  - RNA
- a coat (or shell) made of protein

Vocabulary

- ❒ **DNA**—A chemical that carries instructions for making a new cell; stands for **d**e**o**xyribo**n**ucleic **a**cid.
- ❒ **host**—In this case, a living cell that is the home for a virus.
- ❒ **metabolism**—All of the activities of a cell that allow it to live, grow, mature, and reproduce.
- ❒ **RNA**—A chemical that carries instructions for making a new cell; stands for **r**ibo**n**ucleic **a**cid.

Characteristics

Viruses
- are not cells
- are much smaller than cells
- are active in cells and reproduce there
- take over the metabolism of the host cells

After a virus reproduces in a cell, the cell dies and releases more viruses that can enter other cells.

Infection

- Viruses can infect cells such as
  - animal cells
  - plant cells
  - bacterial cells
- Viruses can cause diseases, such as
  - chicken pox
  - whooping cough
  - measles

Viruses
LOWER CHALLENGE

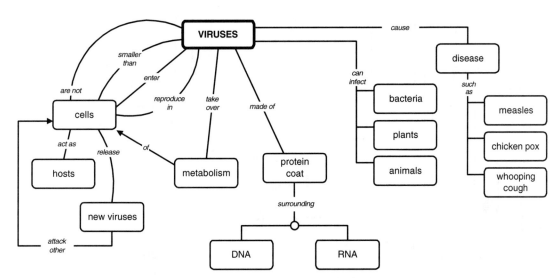

Score: 11 words

Starting hints: The four connectors between the two upper left items should provide a strong clue that the missing word is *cells*. Another good starting point is the pair of boxes with *RNA* on the right.

Notes: Remember that this map notation O means "or."

Viruses
HIGHER CHALLENGE

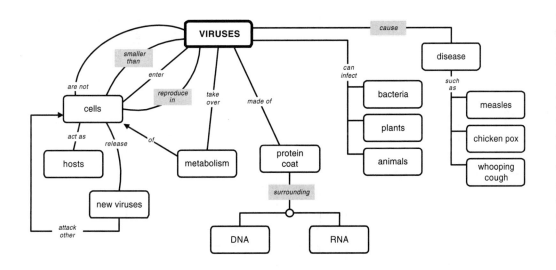

Score: 15 (11 words + 4 connector labels)

Starting hints: The seed item *measles* points to both the item *diseases* and the connector *cause*. The seed item *RNA* is in the core of the virus and has a *protein coat* that *surrounds* it.

Notes: In the higher challenge map, four of the connectors are also missing. The approach here is to complete the map as much as possible with the connectors given, and then determine the missing links or relationships.

Concept Map: Protista

THINKING CONNECTIONS: Life Science Book B — Cell Biology

Name _____ Period _____
Date _____

Directions: Select words from the word list to fill in the blank map items. Use each word only once, and use all the words on the list.

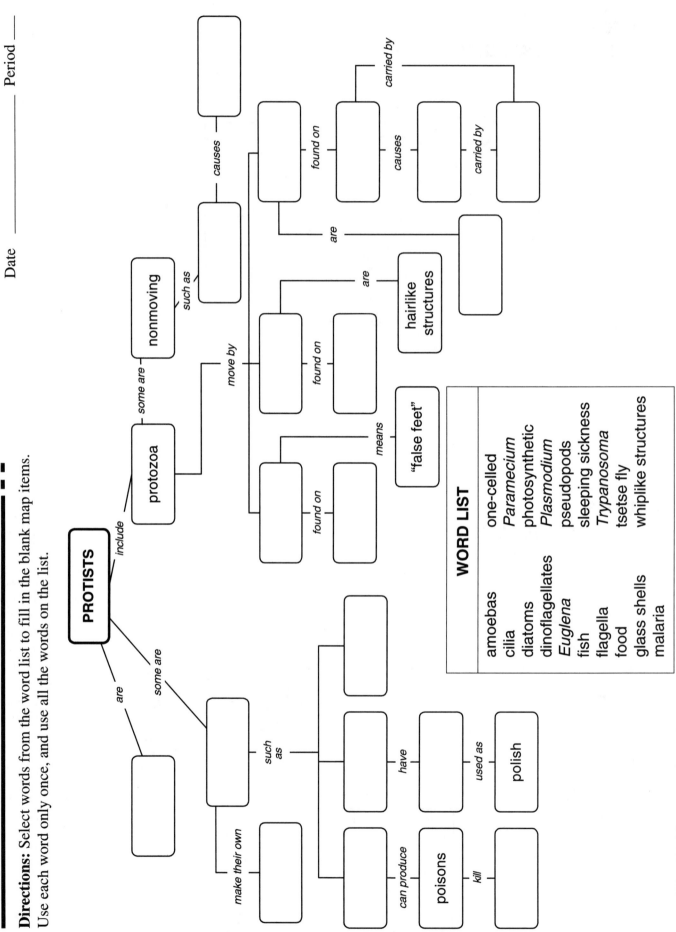

WORD LIST

amoebas	one-celled
cilia	*Paramecium*
diatoms	photosynthetic
dinoflagellates	*Plasmodium*
Euglena	pseudopods
fish	sleeping sickness
flagella	*Trypanosoma*
food	tsetse fly
glass shells	whiplike structures
malaria	

Concept Map: Protista

Directions: Select words from the word list to fill in the blank map items. Use each word only once, and use all the words on the list.

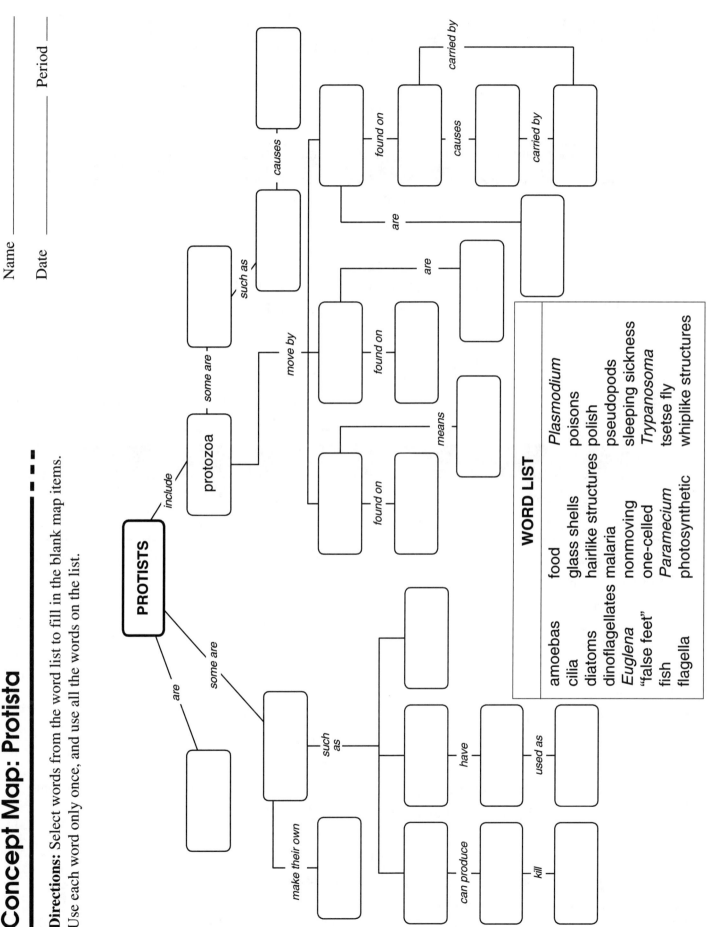

Critical Thinking → CONCEPT FILE

Protista

Background

Protista
- are simple organisms
- are one-celled
- include
  - simple algae
  - protozoa
  - slime molds

Vocabulary

- ❑ **anal pore**—A tiny opening through which wastes leave the cell.
- ❑ **cilia** (singular cilium)—Hairlike structures that create movement by beating in unison.
- ❑ **contractile vacuole**—Stores excess water in the cell; can squeeze water out of the cell.
- ❑ **eyespot**—A cell part sensitive to light.
- ❑ **food vacuole**—A place in the cell where food is digested and stored.
- ❑ **gullet**—The end of the oral groove.
- ❑ **oral groove**—A channel lined with cilia that brings food into *Paramecium*.

Major Groups

Plantlike

Plantlike protists
- have chloroplasts that contain chlorophyll
- make their own food through photosynthesis

Examples
- *Euglena*
  - moves by flagella
  - has red eyespot
- diatoms
  - have glasslike shells of silica
  - accumulate and form "diatomaceous earth"
- dinoflagellates
  - have glasslike shells
  - cause "red tide" (poisonous to fish)

Animal-like

Animal-like protists include the protozoa and are classified by their type of movement:
- flagella (sing. flagellum)
  - are whiplike structures
  - are found on *Trypanosoma* (carried by tsetse fly); cause sleeping sickness
- cilia (singular cilium)
  - are hairlike structures
  - are found on *Paramecium*
- pseudopodia (singular pseudopodium)
  - reference to "false feet" of cytoplasm that flow from one place to another
  - example is *Amoeba*
- nonmotile
  - most are parasitic forms
  - example is *Plasmodium*; the cause of malaria (carried by mosquitoes)

Protista
LOWER AND HIGHER CHALLENGE

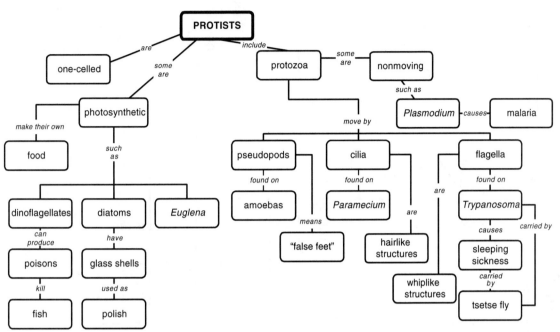

Score:
Lower Challenge 19 words
Higher Challenge 24 words

Starting hints: On the lower challenge map, the seed item *"false feet"* references *pseudopods*. The seed item *hairlike structures* references *cilia*.

On the higher challenge map, the connector *make their own* suggests *food* and establishes the photosynthetic branch of the map. The connector *carried by* on the lower right suggests a vector, in this case the *tsetse fly*.

Concept Map: Sarcodines

Name _____
Date _____ Period _____

Directions: Select words from the word list to fill on the blank map items. Use each word only once, and use all the words on the list.

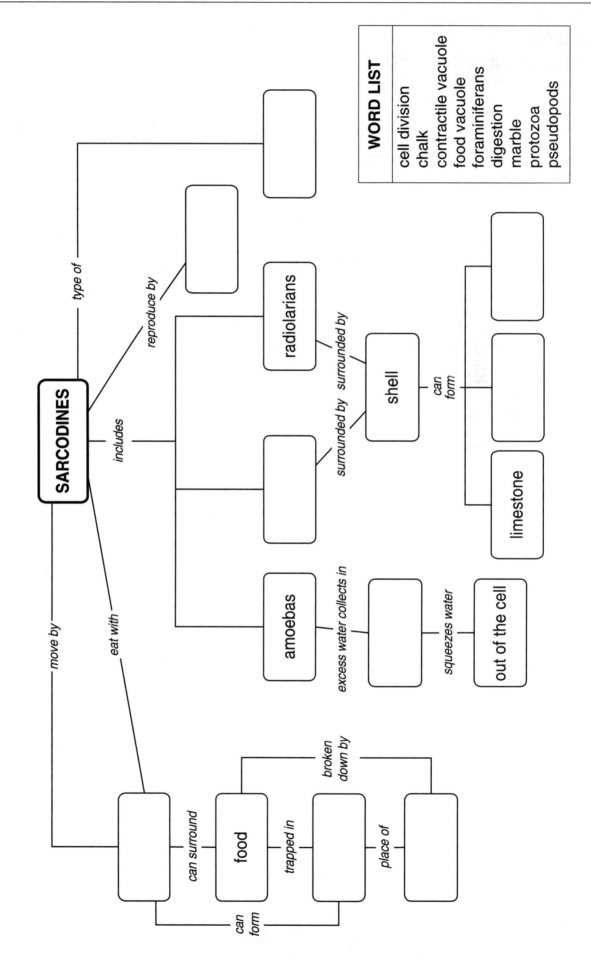

WORD LIST

cell division
chalk
contractile vacuole
food vacuole
foraminiferans
digestion
marble
protozoa
pseudopods

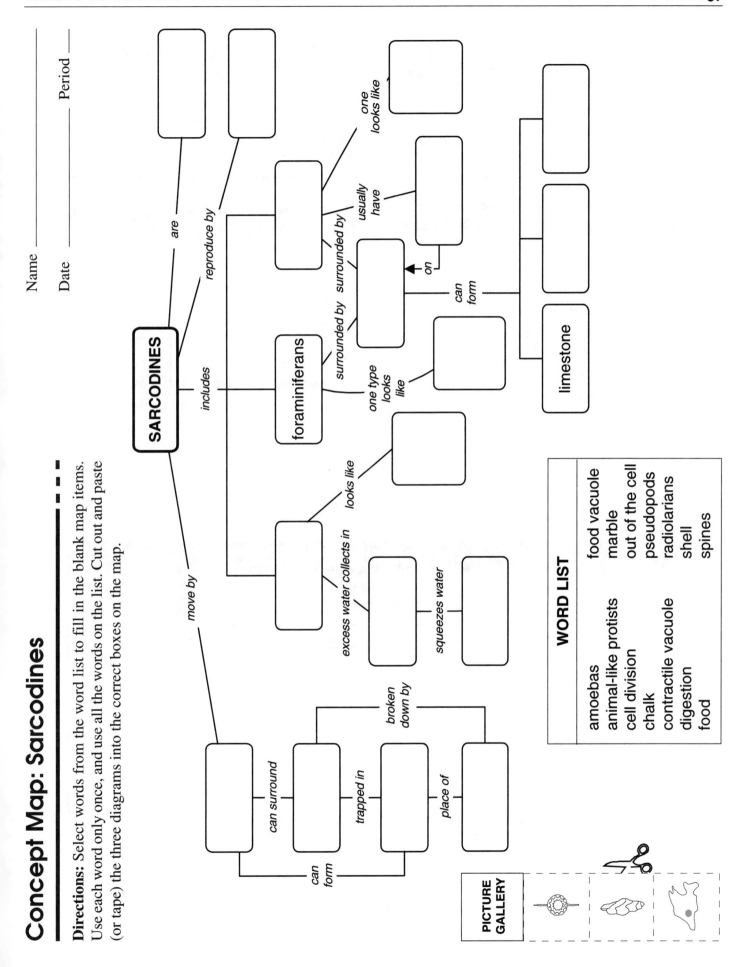

THINKING CONNECTIONS: Life Science Book B   Cell Biology

Critical Thinking

Sarcodines

Background

Sarcodines
- are single-celled organisms
- move by pseudopods
- feed by surrounding food and trapping it in food vacuoles
- are animal-like protists
- use cell division as the main method of reproduction

Vocabulary

- ❐ **contractile vacuole**—Stores excess water in the cell; can squeeze water out of the cell.
- ❐ **food vacuole**—A place in the cell where food is digested and stored.
- ❐ **pseudopod**—A "false foot" of cytoplasm that flows easily and moves the cell; can also flow around food, trap it, and create a food vacuole.

Groups

Foraminiferans

- are surrounded by a glasslike shell, usually *without spines*
- can help to form
  – chalk
  – limestone
  – marble

Radiolarians

- are surrounded by a shell that is *often spiny*
- can help to form
  – chalk
  – limestone
  – marble

Amoeba

- has a bloblike appearance
- has contractile vacuoles

34   © 1998 Critical Thinking Books & Software • www.criticalthinking.com • (800) 458-4849

Sarcodines

LOWER CHALLENGE

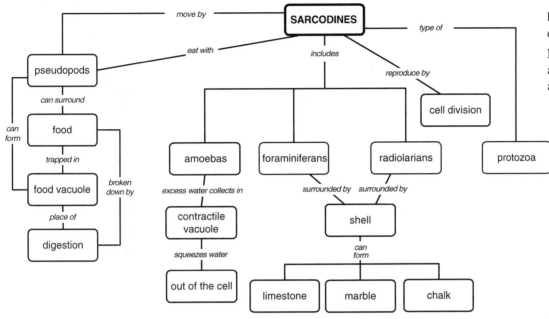

Score: 9 words

Starting hints: The other sarcodine *surrounded by* a *shell* is the *foraminiferans*. The *contractive vacuole* both *collects* and *squeezes* excess water *out of the cell*.

Notes: The terms *marble* and *chalk* can be placed in any order.

HIGHER CHALLENGE

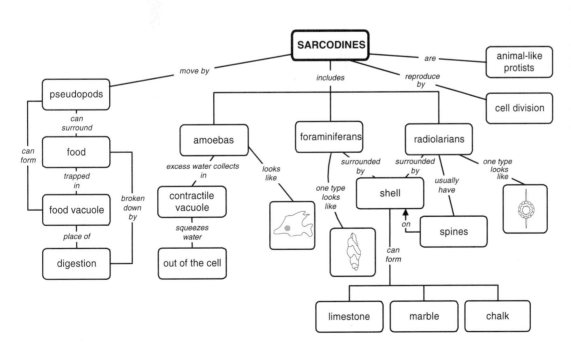

Score: 17 (14 words + 3 pictures)

Starting hints: The seed item *limestone* points to *shell* and suggests its partners are *chalk* and *marble*. The connector *broken down by* suggests *digestion*.

Notes: Point out that there are three pictures for the students to cut out and paste or tape in the correct boxes.

The terms *marble* and *chalk* can be placed in any order.

THINKING CONNECTIONS: Life Science Book B — Cell Biology

Concept Map: Ciliates

Directions: Select words from the word list to fill in the blank map items. Use each word only once, and use all the words on the list.

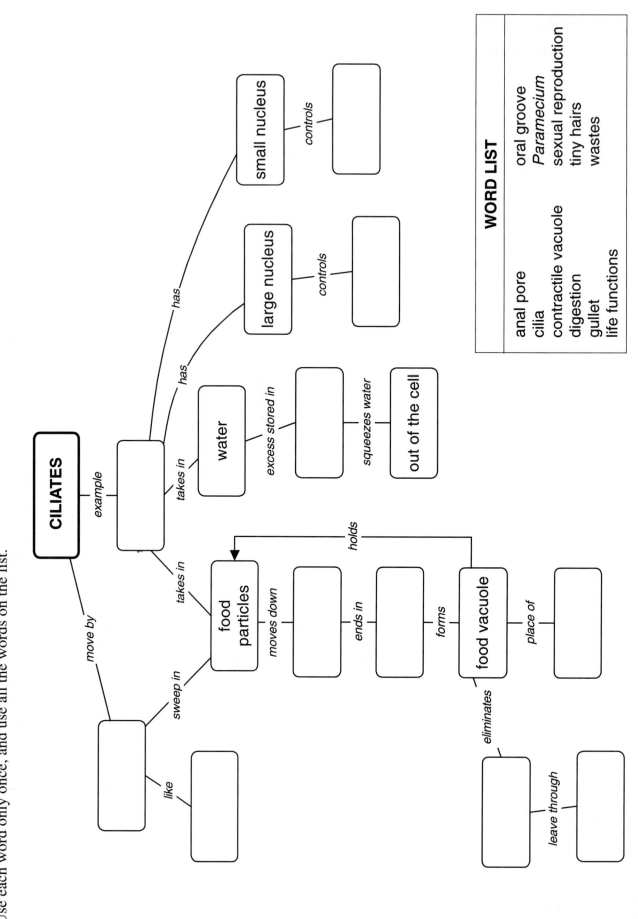

WORD LIST

anal pore
cilia
contractile vacuole
digestion
gullet
life functions
oral groove
Paramecium
sexual reproduction
tiny hairs
wastes

Concept Map: Ciliates

THINKING CONNECTIONS: Life Science Book B — Cell Biology

Name _____
Date _____ Period _____

Directions: Select words from the word list to fill in the blank map items. Use each word only once, and use all the words on the list. Write the letter of each label on the diagram in the box that corresponds to its name.

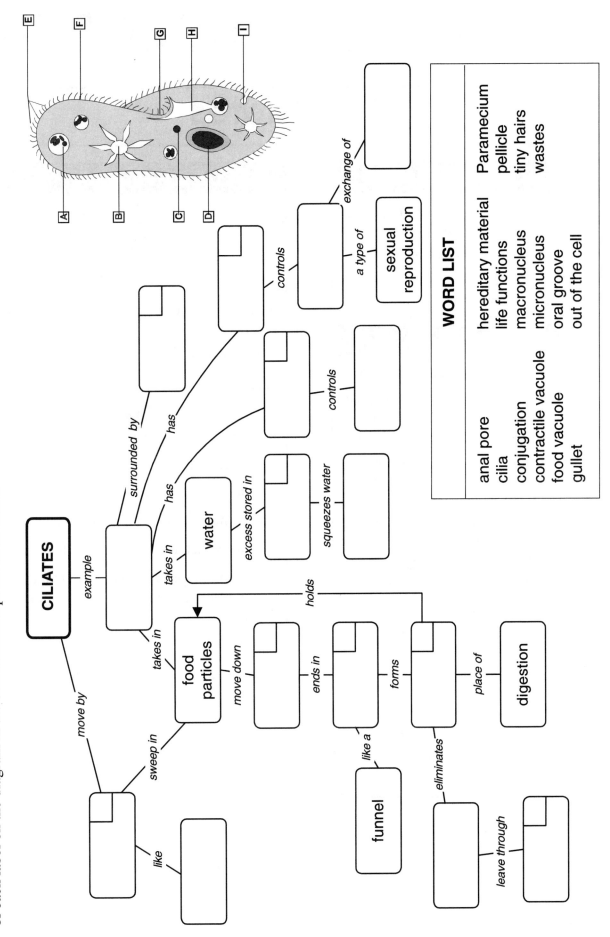

WORD LIST

anal pore	hereditary material
cilia	life functions
conjugation	macronucleus
contractile vacuole	micronucleus
food vacuole	oral groove
gullet	out of the cell
Paramecium	
pellicle	
tiny hairs	
wastes	

Ciliates

Background

Ciliates are one-celled organisms classified as Protista:
- move by cilia (singular: cilium)
- live in water

Vocabulary

- ❏ **cilia**—Small hairlike structures.
- ❏ **conjugation**—A type of sexual reproduction in which hereditary materials are exchanged.
- ❏ **contractile vacuole**—A water-storing area in the cell. Can contract and squeeze water out.
- ❏ **food vacuole**—A place of food storage and digestion.
- ❏ **gullet**—A funnel-like structure at the end of the oral groove.
- ❏ **macronucleus**—A large nucleus that controls cell functions.
- ❏ **micronucleus**—A small nucleus that controls conjugation.
- ❏ **oral groove**—A crease in the side of the *Paramecium*.
- ❏ **pellicle**—A tough, protective outer covering.

Example

Paramecium

Control

- The macronucleus controls life functions of the cell.
- The micronucleus controls conjugation.

Feeding

- Cilia sweep food into the cell.
- Food moves into the oral groove.
- At the gullet, food collects in a food vacuole.
- After digestion, vacuole moves to the edge and eliminates remaining materials through an anal pore.
- Water also enters the cell with food.
- Extra water is stored in a vacuole.
- Vacuole can squeeze extra water out of the cell.

Ciliates

LOWER CHALLENGE

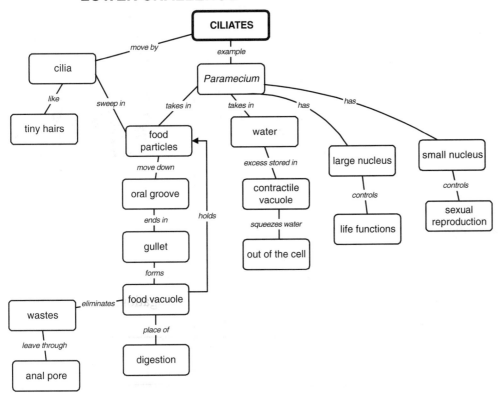

Score: 11 words

Starting hints: The seed item *food vacuole* is the site of *digestion* and is formed by the *gullet*.

HIGHER CHALLENGE

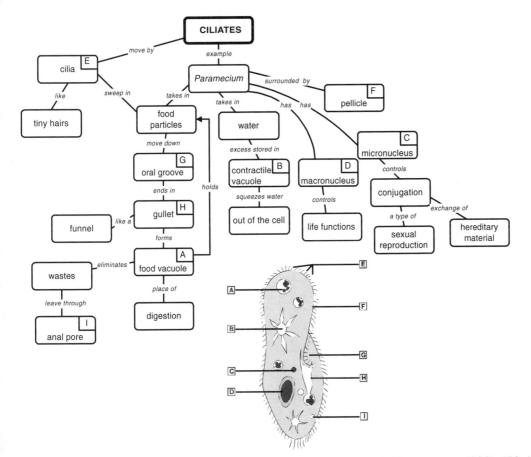

Score: 25 (16 words + 9 letters)

Starting hints: The seed item *digestion* occurs in a structure (*food vacuole*) formed by something that is funnel-like (*gullet*).

Notes: You may want to make the diagram-matching portion optional. When checking maps, make sure that the letters match the words in the boxes, not necessarily the position on the map.

Students may misplace a word on the map but correctly match the word with the diagram.

Diagram Key:
- A food vacuole
- B contractile vacuole
- C micronucleus
- D macronucleus
- E cilia
- F pellicle
- G oral groove
- H gullet
- I anal pore

Concept Map: Flagellates

Directions: Select words from the word list to fill in the blank map items. Use each word only once, and use all the words on the list. Write the letter of each label on the diagram in the box that matches its name.

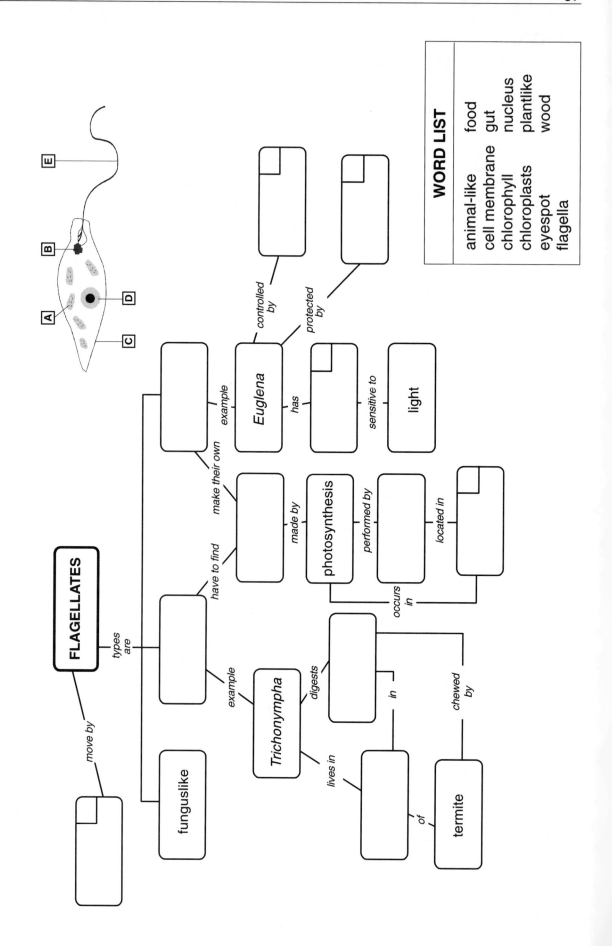

WORD LIST

animal-like
cell membrane
chlorophyll
chloroplasts
eyespot
flagella
food
gut
nucleus
plantlike
wood

THINKING CONNECTIONS: Life Science Book B — Cell Biology

Concept Map: Flagellates

Directions: Select words from the word list to fill in the blank map items. Use each word only once, and use all the words on the list. Write the letter of each label on the diagram in the box that matches its name.

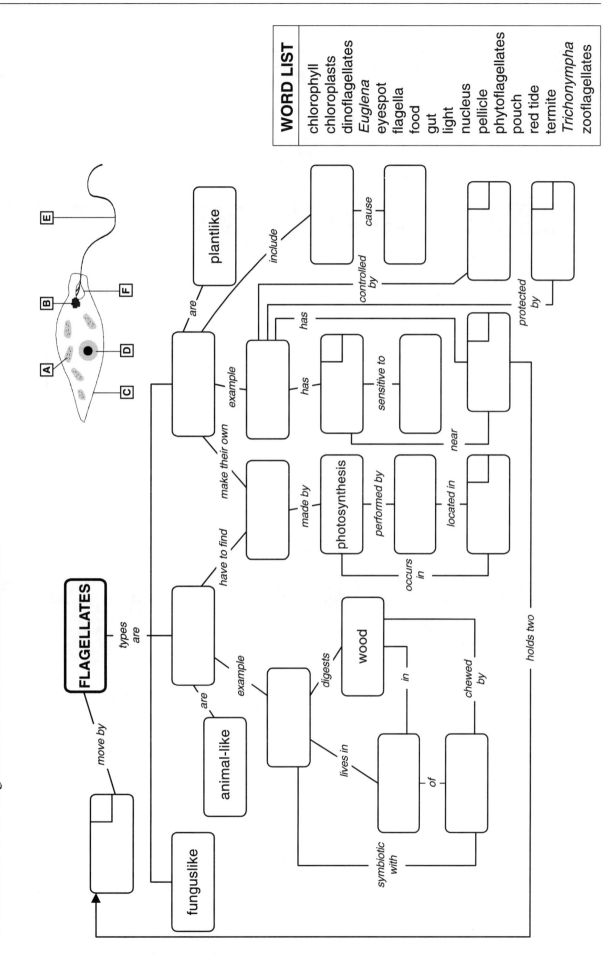

Critical Thinking

Flagellates

Types

Flagellates move by means of one or more flagella, which are thin, hairlike structures.

There are three types of flagellates:

- funguslike
- phytoflagellates (plantlike)
  – make their own food
  – include dinoflagellates (a type that causes red tide)
- zooflagellates (animal-like)
  – get food from the environment

Vocabulary

- ❒ **chlorophyll**—A green pigment used during food manufacture.
- ❒ **chloroplast**—The part of the cell where photosynthesis occurs.
- ❒ **photosynthesis**—The process by which living things make their own food.
- ❒ **red tide**—A growth of dinoflagellates that has a reddish appearance in the water and that can poison other organisms.
- ❒ **symbiosis**—A condition where two organisms live together and benefit each other.

Examples

Trichonympha

- is a zooflagellate
- lives in the gut of termites
- helps the termite digest wood
- is symbiotic with the termite

Euglena

- is a phytoflagellate
- makes its own food in its chloroplasts
- has a red eyespot near the pouch that is sensitive to light
- is protected by an outer pellicle
- is centrally controlled by its nucleus
- has a pouch (can take in food at times)
- has two flagella wrapped around each other

Flagellates

LOWER CHALLENGE

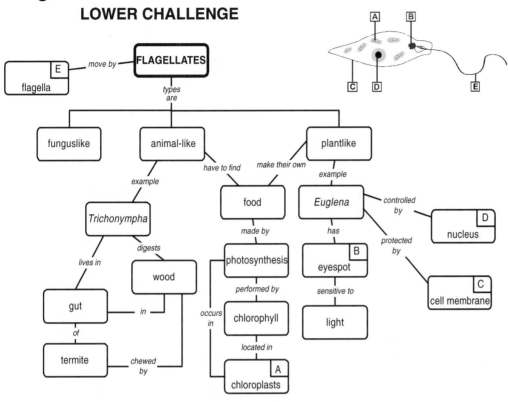

Score: 16 (11 words + 5 letters)

Starting hints: The connector *digests* points to *wood* or *food*, but the specific seed *Trichonympha* strongly suggests *wood*. (The item *food* is used under the connectors *have to find* and *make their own*.)

The *eyespot* is *sensitive to* the seed item *light*.

Notes: You may want to make the diagram-matching portion optional. When checking maps, make sure that the letters match the words in the boxes, not necessarily the position on the map. Students may misplace a word on the map but correctly match the word with the diagram.

Diagram Key:
- A chloroplasts
- B eyespot
- C cell membrane
- D nucleus
- E flagella

HIGHER CHALLENGE

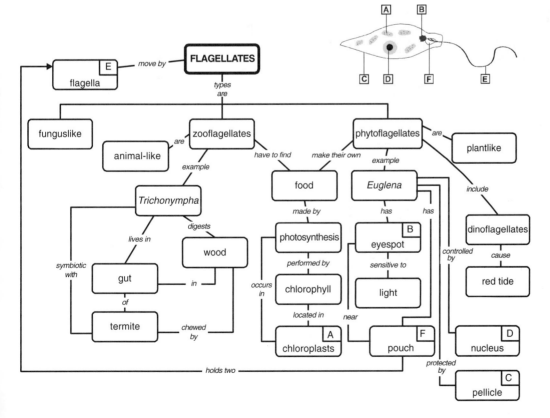

Score: 23 (17 words + 6 letters)

Starting hints: *Food* is *made by* the seed item *photosynthesis*. *Trichonympha* digests the seed item *wood*.

Notes: You may want to make the diagram-matching portion optional. When checking maps, make sure that the letters match the words in the boxes, not necessarily the position on the map. Students may misplace a word on the map but correctly match the word with the diagram.

Diagram Key:
- A chloroplasts
- B eyespot
- C pellicle
- D nucleus
- E flagella
- F pouch

THINKING CONNECTIONS: Life Science Book B

Cell Biology

Name _____
Date _____ Period _____

Concept Map: Sporozoans

Directions: Select words from the word list to fill in the blank map items. Use each word only once, and use all the words on the list.

WORD LIST
- host
- human being
- human liver
- malaria
- mosquito
- new *Plasmodium*
- parasites
- *Plasmodium*
- protista
- spores

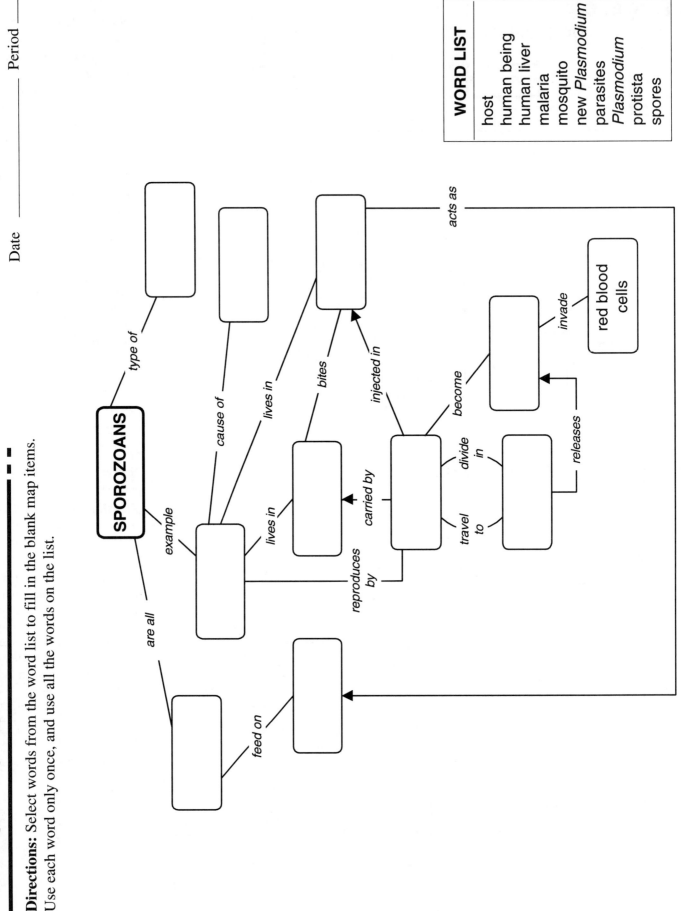

44   © 1998 Critical Thinking Books & Software • www.criticalthinking.com • (800) 458-4849

Concept Map: Sporozoans

THINKING CONNECTIONS: Life Science Book B — Cell Biology

Directions: Select words from the word list to fill in the blank map items. Use each word only once, and use all the words on the list.

WORD LIST
- body fluids
- fever
- host
- human being
- human liver
- malaria
- mosquito
- new *Plasmodium*
- parasites
- *Plasmodium*
- protista
- red blood cells
- saliva

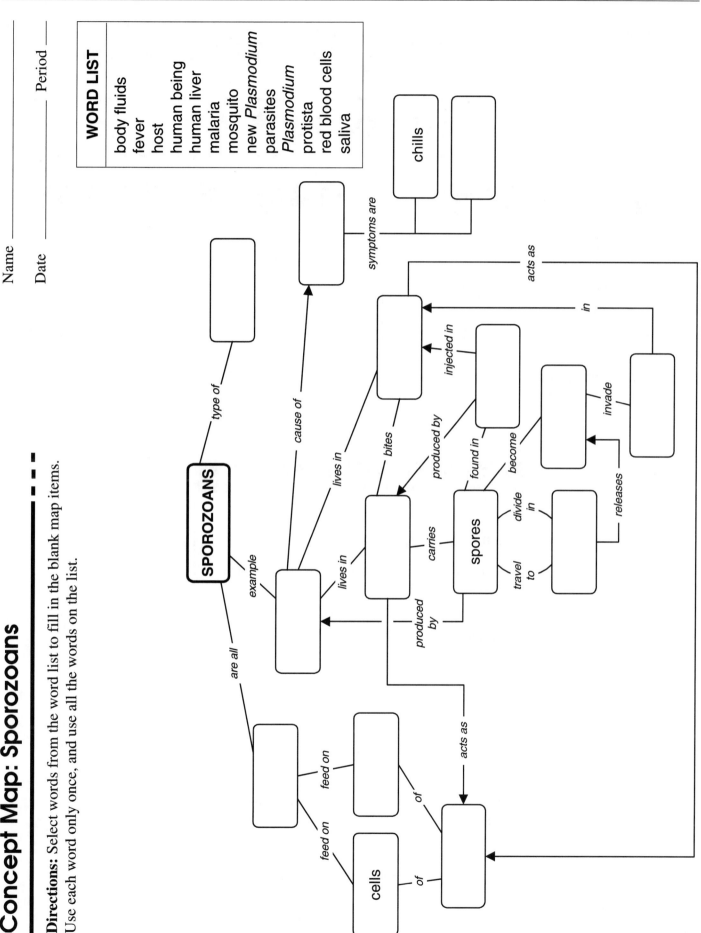

Critical Thinking → Concept File

Sporozoans

Characteristics

The sporozoans
- are one-celled protists
- cannot move on their own
- are all parasitic
  – feed on other cells and body fluids of their hosts

Some sporozoans, like *Plasmodium*, cause diseases in human beings.

Vocabulary

- **parasite**—An organism that gets food from another living organism and harms it in the process.
- **host**—An organism that is a home and a food source for a parasite.

Plasmodium

- This organism causes malaria.
- It lives in and produces spores in the mosquito.
- When a mosquito bites a person, it injects the spores in its saliva into the blood.
- Spores travel to the human liver, where they divide.
- New *Plasmodium* organisms are released from the spores and invade the red blood cells in the human being.
- If a mosquito bites this person, it can pick up more *Plasmodium* organisms in the red blood cells.
- Severe chills and fever are the symptoms of malaria.

THINKING CONNECTIONS: Life Science Book B — Cell Biology

Sporozoans

LOWER CHALLENGE

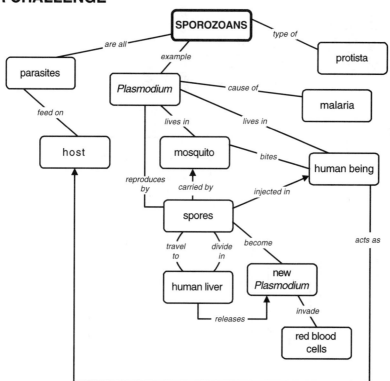

Score: 10 words

Starting hints:

Notes: Point out that the arrows are used to clarify the relationships between items and that students should check these carefully.

HIGHER CHALLENGE

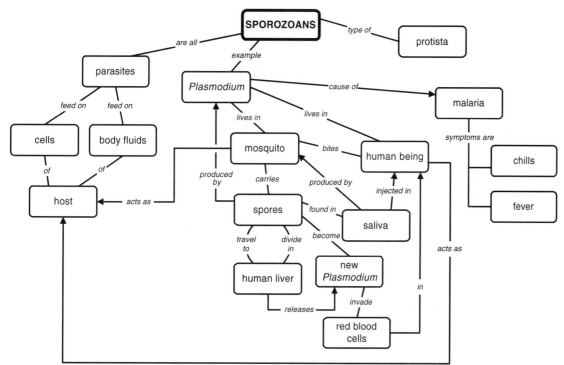

Score: 13 words

Starting hints: The connector *symptoms are* suggests a disease, in this case *malaria*. The connector *example* references the only example in the list, *Plasmodium*.

Notes: Point out that the arrows are used to clarify the relationships between items and that students should check these carefully.

© 1998 Critical Thinking Books & Software • www.criticalthinking.com • (800) 458-4849

Concept Map: Algae

Directions: Select words from the word list and fill in the blank map items. Use each word only once, and use all the words on the list.

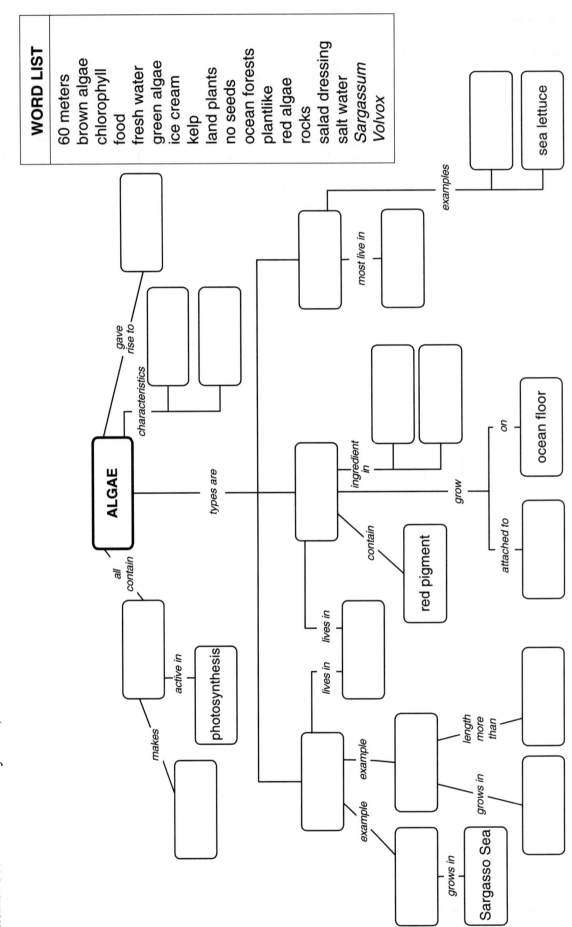

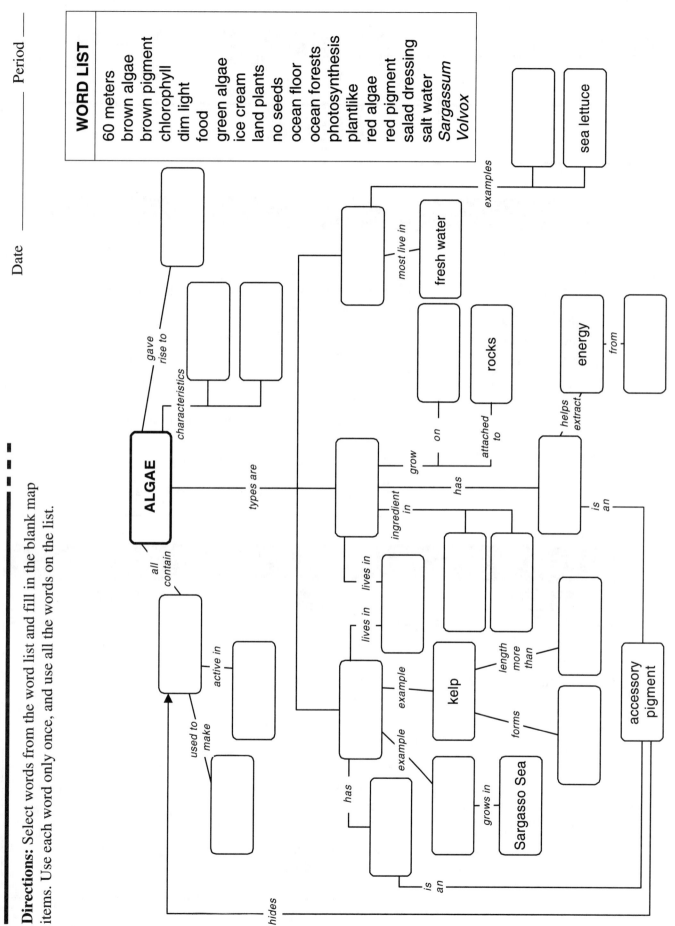

Critical Thinking — CONCEPT FILE

Algae

Background

Algae are simple plants that
- make their own food (photosynthesis)
- have chlorophyll
- do not reproduce by seeds
- are ancestors of land plants

Vocabulary

- **accessory pigment**—A pigment other than chlorophyll, usually hides chlorophyll.
- **chlorophyll**—A green pigment used during food manufacture.
- **kelp**—A type of brown algae, grows in huge ocean "forests."
- **photosynthesis**—The process by which living things make their own food.
- **Sargasso Sea**—An area of the ocean where *Sargassum* grows.
- **Sargassum**—A type of brown algae.
- **Volvox**—A colonial form of green algae.

Three Major Groups

Green Algae

Most live in fresh water.

Examples
- *Volvox*
- sea lettuce

Red Algae

- live in salt water on the ocean floor
- will often attach to rocks
- have red accessory pigment which helps the algae get energy from dim light
- are used in salad dressing and ice cream

Brown Algae

- live in salt water
- have brown accessory pigment

Examples
- *Sargassum*—grows in the Sargasso Sea
- kelp—grows over 60 meters in large ocean forests

Algae

LOWER CHALLENGE

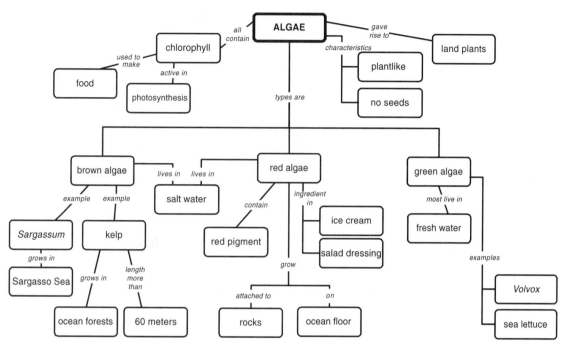

Score: 18 words

Starting hints: The connector *length more than* suggests the item *60 meters*. The seed item *Sargasso Sea* is the site of the growth of *Sargassum*. The seed item *sea lettuce* is an example of *green algae*.

Notes: Even though there is no color legend included with this map, you might suggest to students that they color in the brown, red, and green branches of algae with the appropriate colors. It will make the map easier to read.

HIGHER CHALLENGE

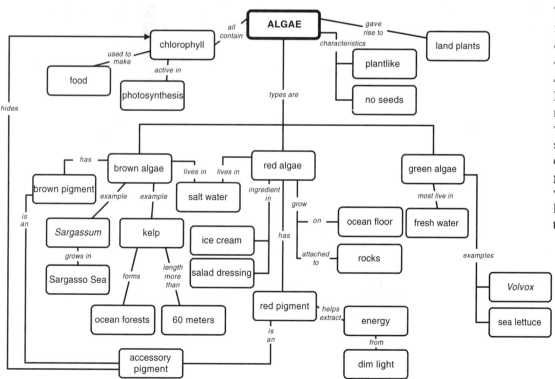

Score: 20 words

Starting hints: The connector *length more than* suggests the item *60 meters*. The seed item *Sargasso Sea* is the site of the growth of *Sargassum*. The seed item *sea lettuce* is an example of *green algae*.

Notes: Even though there is no Color Legend included with this map, you might suggest to students that they color in the brown, red, and green branches of algae with the appropriate colors. It will make the map easier to read.

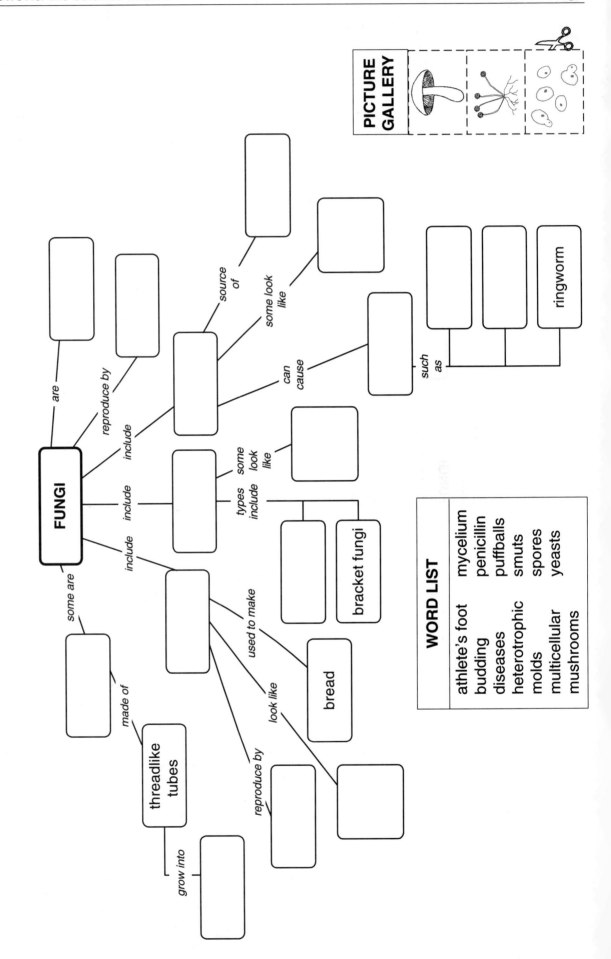

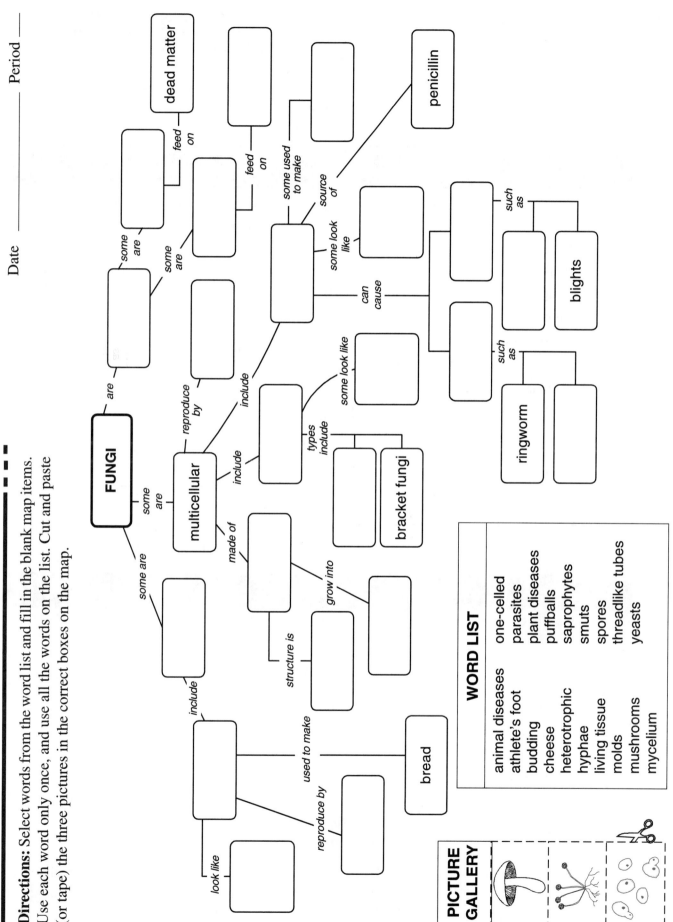

Critical Thinking → Concept File

Fungi

Characteristics

- Fungi are plants that do not make their own food.
- They have no chlorophyll and do not carry on photosynthesis.
- Some fungi
  – are parasitic
  – are saprophytic
  – are unicellular; others are multicellular.
- Multicellular forms are made of thin threadlike tubes called hyphae (singular: hypha).
- The entire mass of hyphae is called the mycelium.
- Multicellular forms reproduce by spores.

Vocabulary

- ❏ **antibiotics**—Chemicals that kill or medically control certain living organisms, e.g., penicillin.
- ❏ **budding**—Reproduction where a young organism grows from the cell of its parent.
- ❏ **chlorophyll**—A green pigment used during food manufacture.
- ❏ **multicellular**—Living organisms consisting of more than one cell.
- ❏ **parasites**—Organisms that feed on living organisms.
- ❏ **photosynthesis**—The process by which living things make their own food.
- ❏ **saprophytes**—Organisms that feed on dead materials.
- ❏ **unicellular**—Living organisms consisting of just one cell.

Types

Yeasts
- are unicellular
- reproduce by budding
- are used in the process of making bread

Mushrooms
- are multicellular
- are made of hyphae

The body is the mycelium, made of many hyphae.

Examples
- bracket fungi (look like shelves on old wood)
- "toadstools"
- puffballs

Molds
- are multicellular
- cause diseases in animals
  – ringworm
  – athlete's foot
- cause diseases in plants
  – blights
  – smuts
- are source of some antibiotics
- are used in the making of cheese

Fungi

LOWER CHALLENGE

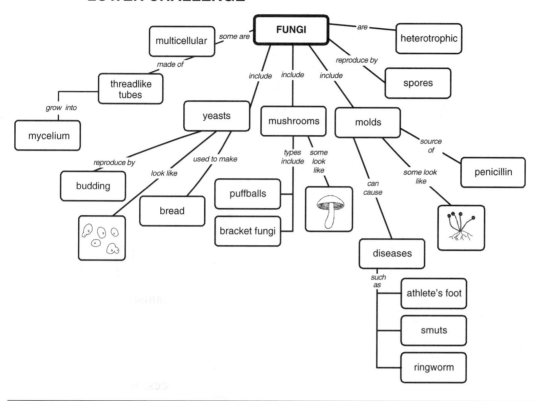

Score: 16 (13 words + 3 pictures)

Starting hints: The connector *can cause* suggests *diseases*. Because the seed item *ringworm* is a disease, the other items in its group are probably diseases also.

Notes: Students will need scissors and tape or glue for this map.

Check pictures carefully. Students may have placed the words incorrectly but then placed the pictures correctly based on the context of their maps.

HIGHER CHALLENGE

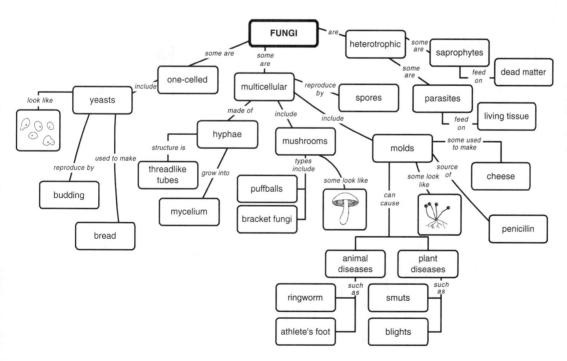

Score: 22 (19 words + 3 pictures)

Starting hints: *Yeast* is the fungus *used to make* the seed item *bread*. The *source* of the seed item *penicillin* is *molds*.

Notes: Check pictures carefully. Students may have placed the words incorrectly but then placed the pictures correctly based on the context of their maps.

Concept Map: Mosses

Directions: Select words from the word list and fill in the blank map items. Use each word only once, and use all the words on the list.

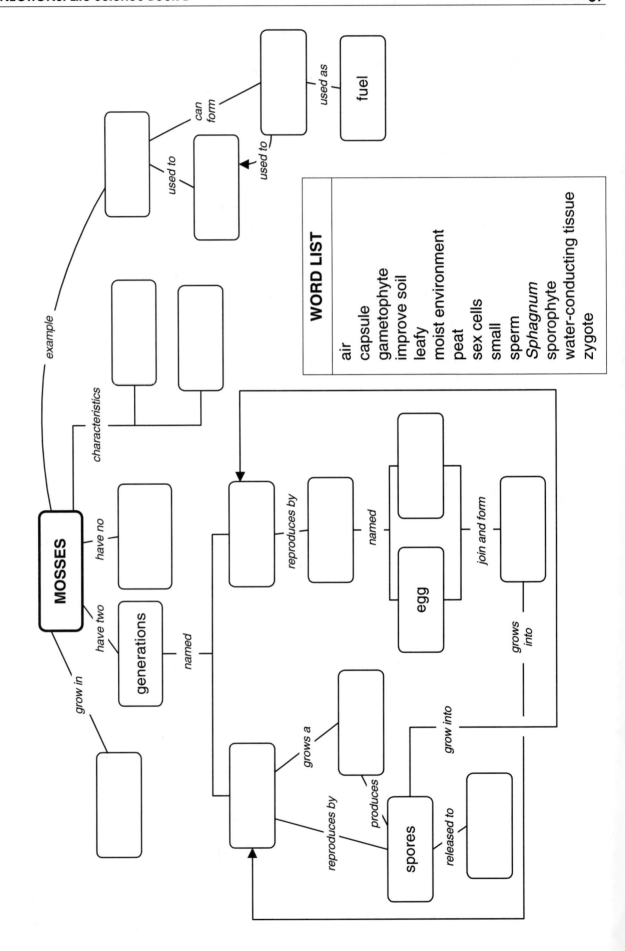

Concept Map: Mosses

THINKING CONNECTIONS: Life Science Book B — Plant Biology

Name _____
Date _____ Period _____

Directions: Select words from the word list and fill in the blank map items. Use each word only once, and use all the words on the list. Then use two different highlighters, colored pencils, or crayons to color in items that are (1) part of the asexual generation and (2) part of the sexual generation. Show your color scheme in the legend.

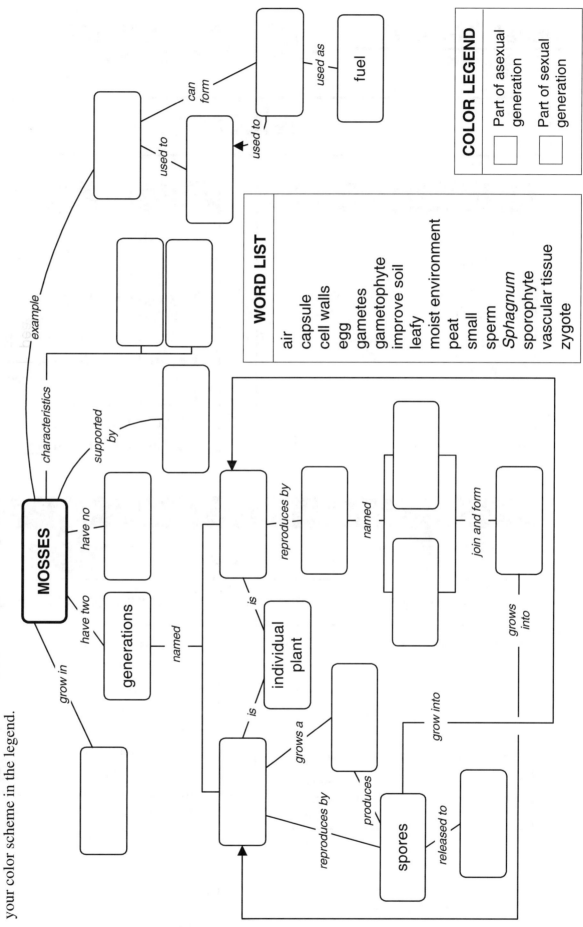

© 1998 Critical Thinking Books & Software • www.criticalthinking.com • (800) 458-4849

THINKING CONNECTIONS: Life Science Book B   Plant Biology

Critical Thinking →

Mosses

Characteristics

Mosses

- are non-vascular plants
- get their support from the cell walls of each cell
- require a moist environment
- are leafy

Vocabulary

- ❏ **gametophyte**—The generation of a plant that reproduces by gametes (sperm and egg).
- ❏ ***Sphagnum***—A type of moss used for improving soil; dead *Sphagnum* plants are known as *peat* moss.
- ❏ **sporophyte**—The generation of a plant that reproduces by spores
- ❏ **vascular tissue**—Tissue that can efficiently move water and food.
- ❏ **zygote**—A fertilized egg.

Alternation of Generations

Sporophyte Generation

- The moss sporophyte is an individual brown, leafless plant.
- The sporophyte reproduces by spores.
- The spores form in a capsule at the top of the plant.
- Spores are released into the air.
- Those spores that land in suitable habitat grow into gametophytes.

Gametophyte Generation

- The moss gametophyte is an individual small, green, leafy plant.
- The gametophyte produces both sperm and egg (gametes).
- Sperm are released and swim in the moisture on the plant.
- If a sperm swims to an egg at the top of the plant and fertilizes it, a zygote is formed.
- The new plant grows out of the top of the gametophyte.
- The new plant is a sporophyte.

Mosses

LOWER CHALLENGE

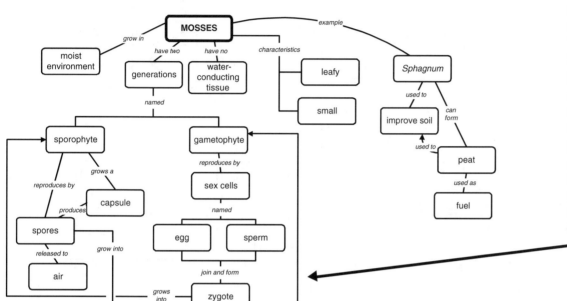

Score: 14 words

Starting hints: The *sporophyte* reproduces by the seed item *spores*. The seed item *egg* is one of the *sex cells*, as is its pair the *sperm*. The seed item *fuel* references *peat*.

Notes: The lower-challenge map uses the term *water-conducting tissue* in place of *vascular tissue*.

On both maps, the section on the lower left of the map is similar to the Alternation of Generations in plants above mosses (i.e., ferns, conifers, flowering plants).

HIGHER CHALLENGE

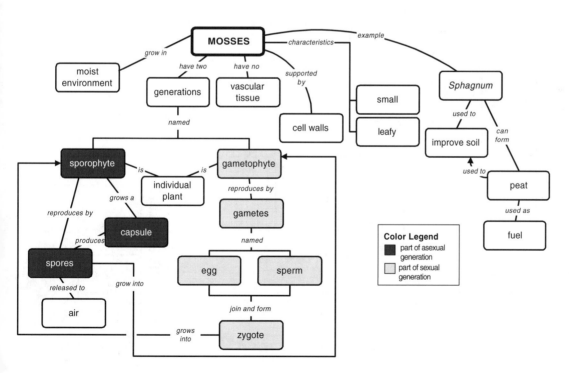

Score: 24 words (16 words + 8 colored items)

Starting hints: The connector label *join and form* is a good place because only two items on the list actually join (sperm and egg). This connection establishes the "sexual generation" branch of the map.

Notes: The higher-challenge map includes a Color Legend and two additional terms. The term *gamete* is used instead of the more general term *sex cells*.

THINKING CONNECTIONS: Life Science Book B — Plant Biology

Concept Map: Vascular Plants

Name _____ Period _____
Date _____

Directions: Select words from the word list and fill in the blank map items. Use each word only once, and use all the words on the list. Cut and paste (or tape) the pictures in the correct boxes on the map.

WORD LIST
- cedar
- evergreen
- flowering plants
- needles
- phloem
- seeds
- vascular tissue
- xylem

Concept map with the following labeled relationships:

- **VASCULAR PLANTS** *have* → [ ] (vascular tissue)
  - *moves* water *up through* → [ ] (xylem) — *type of* → back to vascular tissue
  - *moves* food *down through* → [ ] (phloem) — *type of* → back to vascular tissue
- **VASCULAR PLANTS** *types are*:
  - [ ] (flowering plants) *produce* → [ ] *develop into* → **fruits**
    - *examples* → [ ] [ ] [ ] [ ]
    - fruits *completely enclose* → [ ] (seeds)
  - **cone-bearing plants** *produce* → [ ] (cones) — *protect* → seeds
    - *modified leaves are* → [ ] (needles)
    - *examples* → **spruce**, [ ], [ ]
    - *many are* → [ ] (evergreen) — (cedar)

PICTURE GALLERY
- flowers
- acorn
- pine
- peanuts
- ferns
- pepper
- cones
- banana

Concept Map: Vascular Plants

THINKING CONNECTIONS: Life Science Book B — **Plant Biology**

Name _____ Date _____ Period _____

Directions: Select words from the word list and fill in the blank map items. Use each word only once, and use all the words on the list. Cut and paste (or tape) the pictures in the correct boxes on the map.

WORD LIST
- angiosperms
- cedar
- evergreen
- food
- fruits
- phloem
- seeds
- vascular tissue
- water
- xylem

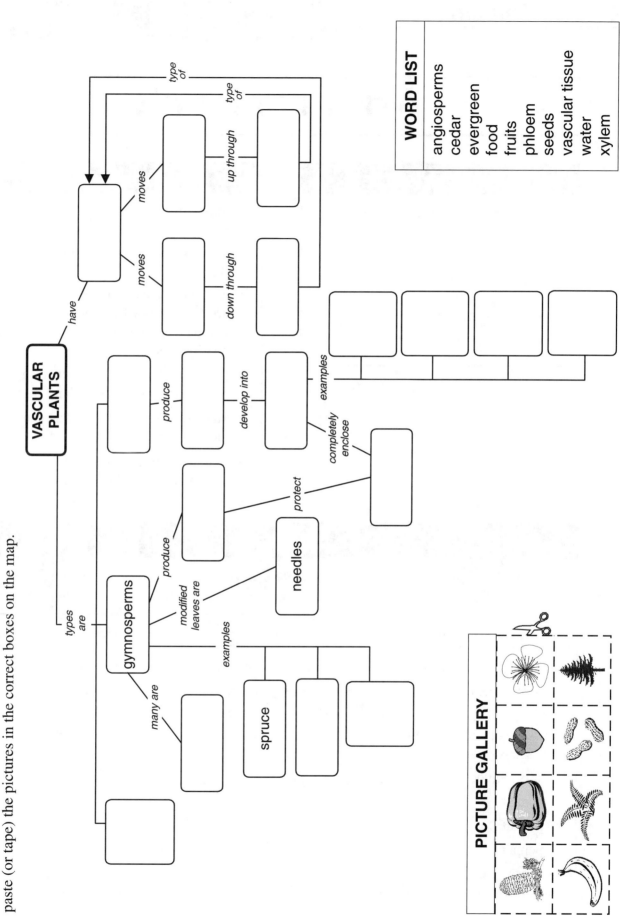

© 1998 Critical Thinking Books & Software • www.criticalthinking.com • (800) 458-4849   61

THINKING CONNECTIONS: Life Science Book B — Plant Biology

Critical Thinking

 CONCEPT FILE

Vascular Plants

Characteristics

The vascular plants have vascular tissues:
- xylem
- phloem

Vocabulary

- **angiosperms**—Means "covered seeds," plants where the seeds are completely covered by the fruit; include all flowering plants.
- **gymnosperms**—Means "exposed seeds," plants where the seeds are not completely enclosed; include all cone-bearing plants.
- **phloem**—Vascular tissue that carries food down through the plant.
- **vascular tissue**—Tissue that can efficiently move water and food.
- **xylem**—Vascular tissue that carries water and minerals up through the plant.

Types

Angiosperms

- All are flowering plants.
- Flower parts develop into fruit.
- Fruit completely encloses the seeds.
- Fruits include
  – banana
  – peanut
  – pepper
  – acorn

Ferns

- Do not produce seeds
- Are usually small, leafy plants

Gymnosperms

- Many are evergreen.
- Needles are modified leaves.
- Most produce seeds in cones, which protect the seeds.
- Examples include
  – cedar
  – spruce
  – pine

Vascular Plants

LOWER CHALLENGE

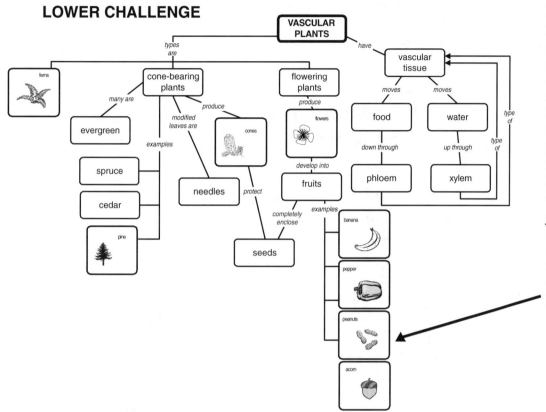

Score: 16 (8 words + 8 pictures)

Starting hints: The seed item *fruits* completely enclose *seeds*. The seed item *cone-bearing plants* have leaves modified into *needles*.

Notes: Each picture on the lower-challenge map is labeled.

This map uses the terms *flowering plants* and *cone-bearing plants* instead of angiosperms and gymnosperms.

The examples of fruit are interchangeable.

HIGHER CHALLENGE

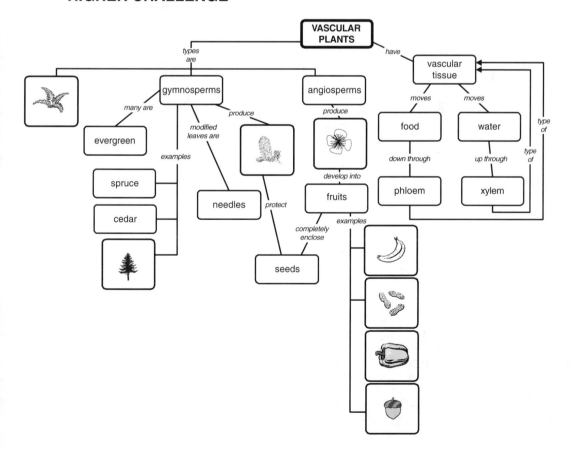

Score: 18 (10 words + 8 pictures)

Starting hints: *Cedar* and the picture of a pine are other *examples* of gymnosperms. The connector *completely enclose* strongly suggests *seeds*, and the term *completely* points to *fruits*.

Notes: The pictures on the higher-challenge map are not labeled.

The higher-challenge map uses the terms *angiosperms* and *gymnosperms*.

Some gymnosperms do not have needles, but many gymnosperms do, and this generalization is an important statement of plant adaptation.

The examples of fruit are interchangeable.

THINKING CONNECTIONS: Life Science Book B — Plant Biology

Concept Map: Ferns

Name _____

Date _____ Period ___

Directions: Select words from the word list to fill in the blank map items. Use each word only once, and use all the words on the list.

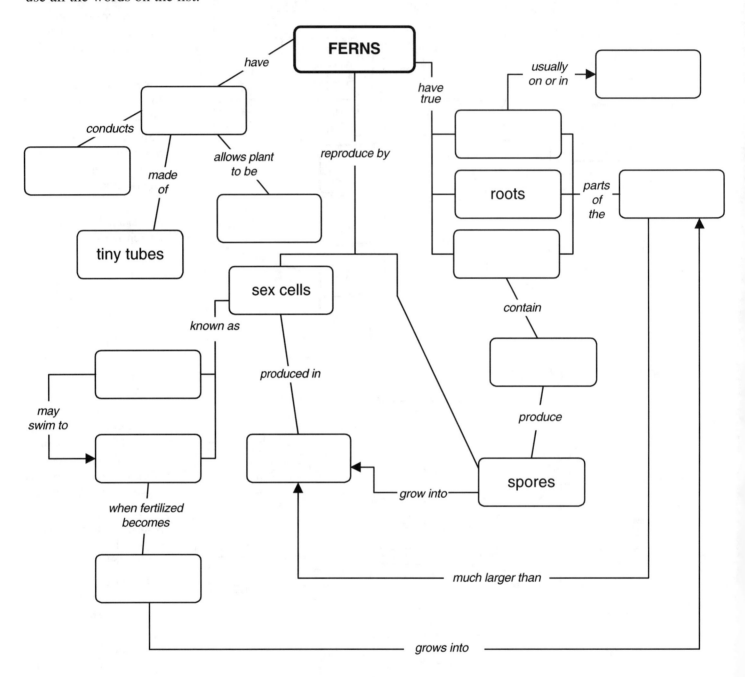

WORD LIST		
conducting tissue	leaves	stems
eggs	sori	tall
gametophyte	sperm	water
ground	sporophyte	zygote

© 1998 Critical Thinking Books & Software • www.criticalthinking.com • (800) 458-4849

THINKING CONNECTIONS: Life Science Book B   Plant Biology

Concept Map: Ferns

Name _____

Date _____ Period ____

Directions: Select words from the word list to fill in the blank map items. Use each word only once, and use all the words on the list. Then use two different highlighters, colored pencils, or crayons to color in items that are (1) part of the asexual generation and (2) part of the sexual generation. Show your color scheme in the legend.

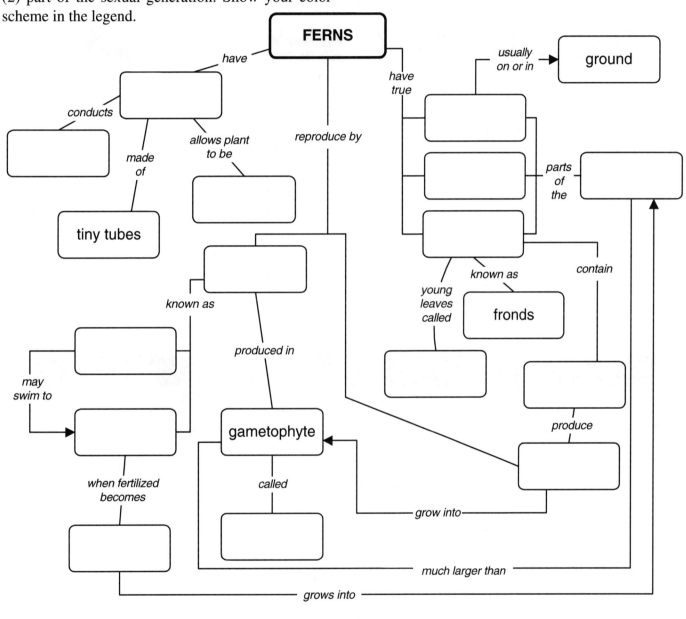

WORD LIST		
egg	roots	stems
fiddleheads	sori	tall
gametes	sperm	vascular tissue
leaves	spores	water
prothallus	sporophyte	zygote

COLOR LEGEND

☐ Part of asexual generation

☐ Part of sexual generation

© 1998 Critical Thinking Books & Software • www.criticalthinking.com • (800) 458-4849   65

THINKING CONNECTIONS: Life Science Book B   Plant Biology

Critical Thinking

Ferns

Characteristics

Ferns have vascular tissues:
- allows for tall growth
- can conduct water to stem and leaves

Ferns have true plant organs:
- roots
- stems
- leaves

Vocabulary

- ❑ **fiddlehead**—A leaf that is unfolding (looks like the top of a violin).
- ❑ **frond**—The leaf of a fern.
- ❑ **gametophyte**—The generation of a plant that reproduces by spores.
- ❑ **sporophyte**—The generation of a plant that reproduces by spores.
- ❑ **vascular tissue**—Tissue that can efficiently move water.
- ❑ **zygote**—A fertilized egg.

Alternation of Generations

Sporophyte Generation

- The sporophyte is the familiar fern plant with its leaves, stems, and roots.
- The sporophyte reproduces by spores.
- Spores are produced in structures on the leaf called sori (singular: sorus)
- Spores are released into the air.
- Those spores that land in suitable habitat grow into gametophytes.

Gametophyte Generation

- The gametophyte is a very small heart-shaped plant.
- A fern gametophyte is called a prothallus.
- The prothallus produces both egg and sperm (gametes).
- Sperm are released and swim in the moisture on the prothallus.
- If a sperm swims to an egg and fertilizes it, a zygote is formed.
- The new plant grows out of the gametophyte.
- The new plant is a sporophyte.

Ferns

LOWER CHALLENGE

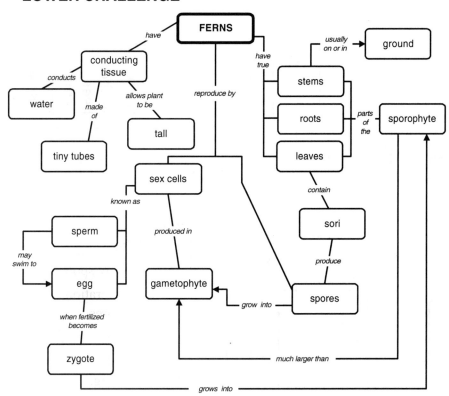

Score: 12 words

Starting hints: The seed item *spores* are *produced* by *sori*. The connector *when fertilized becomes* strongly suggests the item *zygote*. The seed item *tiny tubes* describes the item *conducting tissue*.

Notes: This map covers the basic characteristics and life cycle of ferns.

HIGHER CHALLENGE

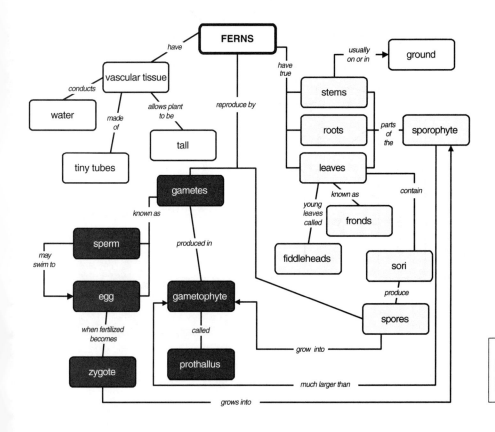

Score: 29 (15 words + 14 colored items)

Starting hints: The seed item *gametophyte* is called a *prothallus* in the ferns. The seed item *fronds* is a name given to fern *leaves*, and young ones are called *fiddleheads*. The seed item *tiny tubes* describes the item *vascular tissue*.

Notes: The higher-challenge map includes additional terms and requires students to color code the stages of alternation of generation.

Some students may choose to color the vascular tissue branch of the map as part of the asexual generation, arguing that vascular tissue allows the sporophyte to attain its height. This is a reasonable argument, but students who leave the branch blank should not lose points.

Concept Map: Roots

Directions: Select words from the word list to fill in the blank map items. Use each word only once, and use all the words on the list.

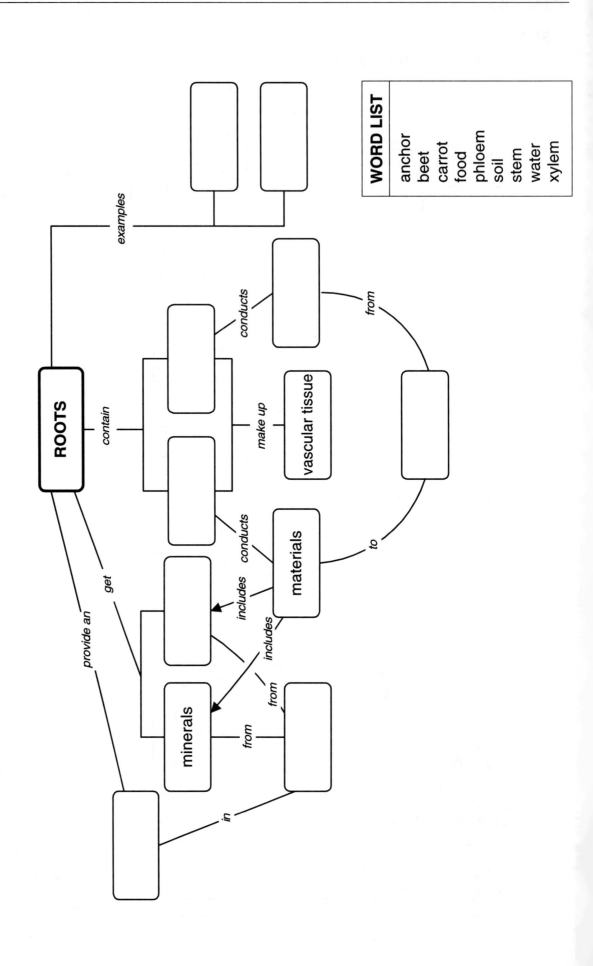

Concept Map: Roots

THINKING CONNECTIONS: Life Science Book B — Plant Biology

Name _____
Date _____ Period _____

Directions: Select words from the word list to fill in the blank map items. Use each word only once, and use all the words on the list.

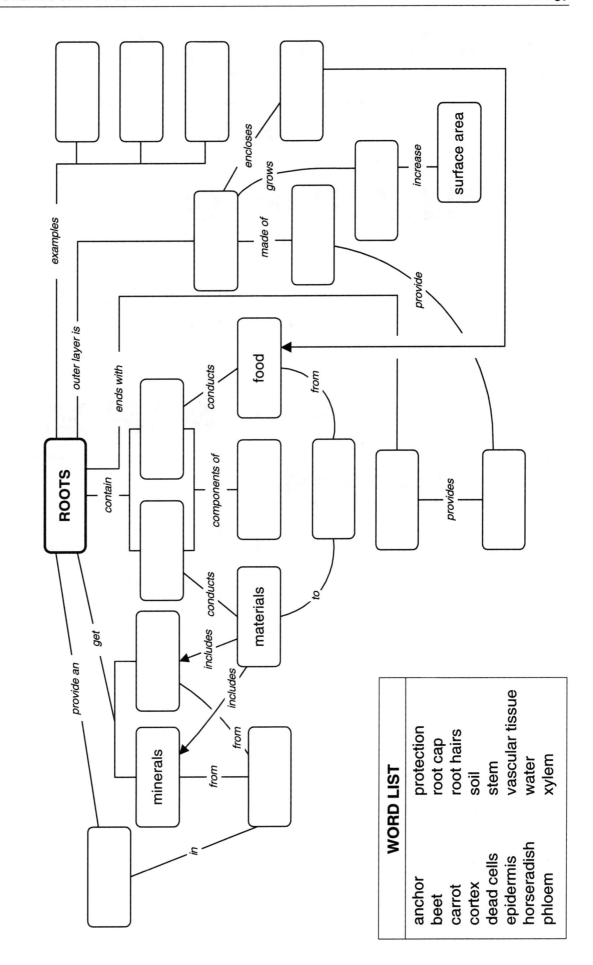

WORD LIST

anchor	protection
beet	root cap
carrot	root hairs
cortex	soil
dead cells	stem
epidermis	vascular tissue
horseradish	water
phloem	xylem

© 1998 Critical Thinking Books & Software • www.criticalthinking.com • (800) 458-4849

Critical Thinking

Roots

Functions

- The root absorbs water and minerals and sends them up to the rest of the plant.
  - xylem moves water and minerals up
- The root stores food produced by the rest of the plant.
  - phloem moves food down into the root
- The root anchors the plant in the soil.

Examples of Roots

Some roots are edible:
- horseradish
- beet
- carrot

Vocabulary

- ❑ **phloem**—Vascular tissue that carries food in plants.
- ❑ **vascular tissue**—Tissue that can efficiently move water and food.
- ❑ **xylem**—Vascular tissue that carries water and minerals in plants.

Structures

- The outside of the root is the epidermis, made of dead cells, which protects the root.

- The tip of the root is covered with the root cap, which also provides protection.

- Some roots have root hairs, which increase the surface areas of the root and allow for more absorption of water.

- The inner part of the root is the cortex, where food is stored.

Roots

LOWER CHALLENGE

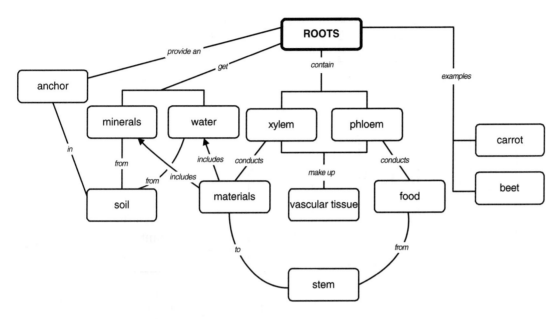

Score: 9 words

Starting hints: The seed item *vascular tissue* is made of *xylem* and *phloem*. Because *water* is closely associated with the seed item *minerals*, *xylem* must be on the left and *phloem* must be on the right.

Notes: This map stresses the basics of root function and parts.

HIGHER CHALLENGE

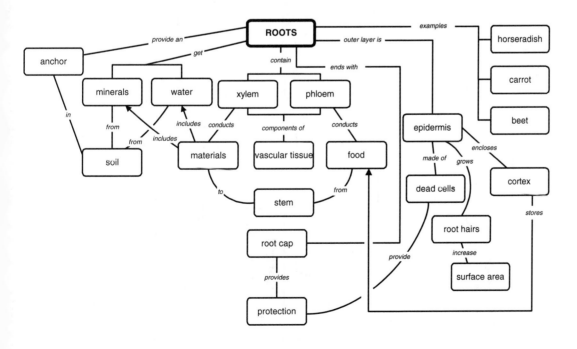

Score: 16 words

Starting hints: *Phloem* conducts the seed item *food*. *Root hairs* will increase the seed item *surface area*.

Notes: The higher-challenge map is conceptually richer with more concepts and many cross connections. Give students adequate time to trace the connections and sort out the concepts.

Concept Map: Stems

Directions: Select words from the word list and fill in the blank map items. Use each word only once, and use all the words on the list.

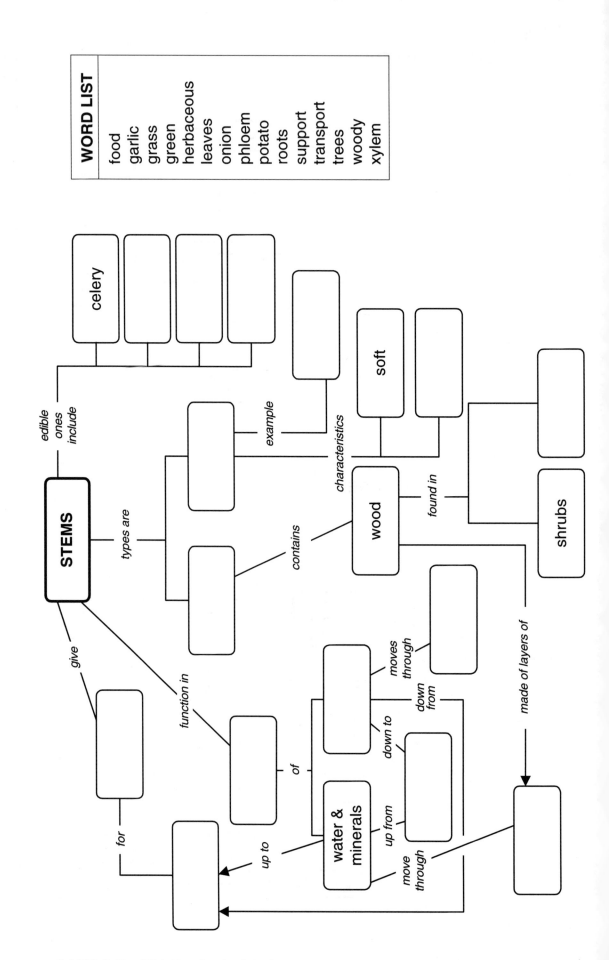

WORD LIST: food, garlic, grass, green, herbaceous, leaves, onion, phloem, potato, roots, support, transport, trees, woody, xylem

Concept Map: Stems

THINKING CONNECTIONS: Life Science Book B — Plant Biology

Name _____
Date _____ Period _____

Directions: Select words from the word list and fill in the blank map items. Use each word only once, and use all the words on the list.

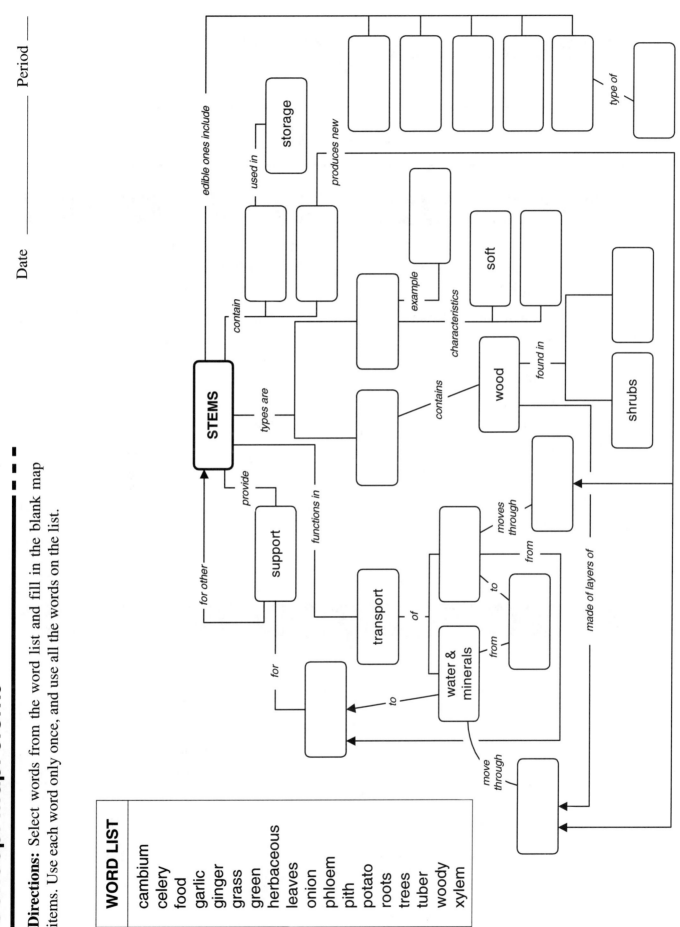

WORD LIST

cambium
celery
food
garlic
ginger
grass
green
herbaceous
leaves
onion
phloem
pith
potato
roots
trees
tuber
woody
xylem

THINKING CONNECTIONS: Life Science Book B — Plant Biology

Critical Thinking  CONCEPT FILE

Stems

Characteristics

Stems
- support the leaves and other stems
- contain xylem
  – tissue that transports water and minerals from the roots to the leaves
- contain phloem
  – tissue that moves food from the leaves to the roots
- contain pith
  – tissue that stores food and water
- contain cambium
  – tissue that makes new xylem and phloem

Vocabulary

Check your understanding—these terms are explained on this page.
- ☐ cambium
- ☐ herbaceous
- ☐ phloem
- ☐ pith
- ☐ tuber
- ☐ xylem

Types of Stems

Herbaceous stems
- are green and soft

Example
- grass stem

Woody stems
- are found in shrubs and trees
- contain wood
- are formed from layers of xylem

Examples of Stems
- Garlic
- Ginger
- Tubers
  – for example, potato
- Onion
- Celery

Stems

LOWER CHALLENGE

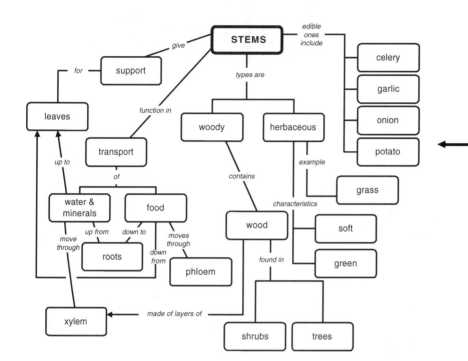

Score: 15 words

Starting hints: The seed item *wood* is found in *trees* and is made of layers of *xylem*. The seed item *water and minerals* move up from the *roots*.

Notes: This map shows the role of stems, their types and a few examples.

The examples here can be in any order.

The part of the map showing the movement of water and food is somewhat complex, but it's important that students grasp how roots, stems, and leaves are all interrelated.

HIGHER CHALLENGE

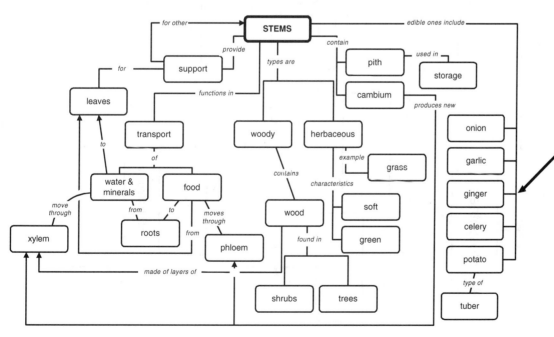

Score: 18 words

Starting hints: The seed item *wood* is found in *trees* and is made of layers of *xylem*. The seed item *water and minerals* move up from the *roots*.

Notes: The higher-challenge map includes more stem functions and structures with fewer connector labels.

The examples of edible stems can be in any order except for *potato*, as it is the only stem identified by type (tuber).

THINKING CONNECTIONS: Life Science Book B Plant Biology

Concept Map: Leaves

Name _____
Date _____ Period _____

Directions: Select words from the word list and fill in the blank map items. Use each word only once, and use all the words on the list.

WORD LIST

carbon dioxide
chlorophyll
cuticle
epidermis
food
guard cells
lettuce
oxygen
parsley
phloem
sugar
sun's energy
spinach
water
vein
xylem

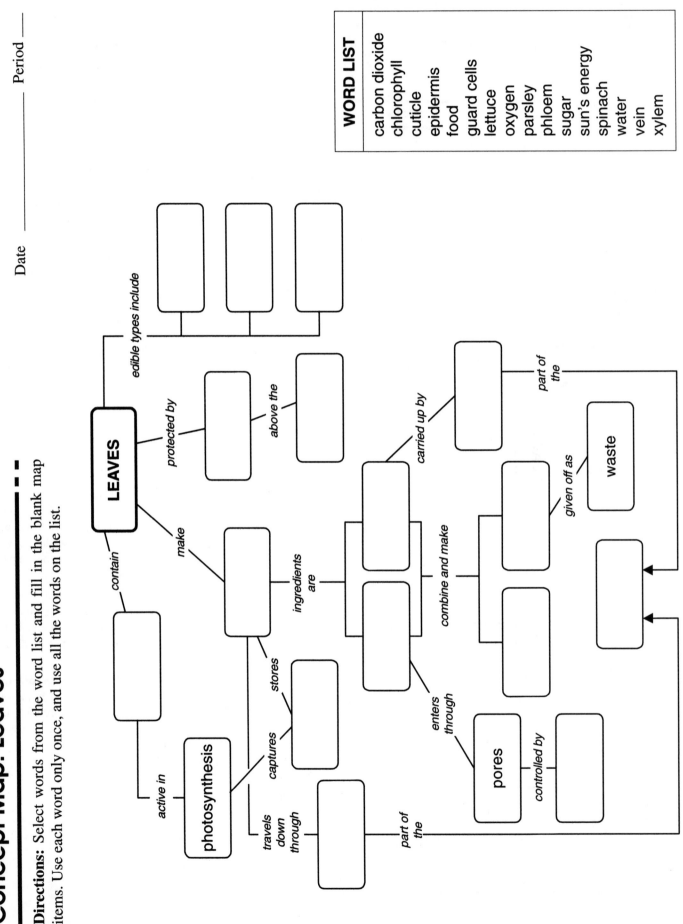

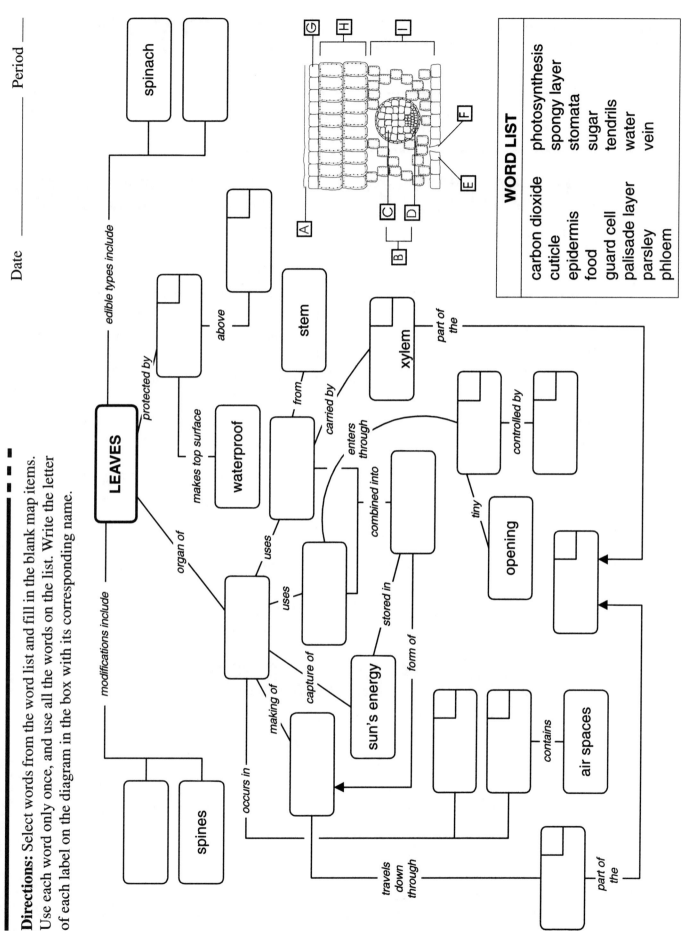

THINKING CONNECTIONS: Life Science Book B — Plant Biology

Critical Thinking

Leaves

General Information

A leaf is the organ of photosynthesis. Some leaves are modified into other structures:
- spines
- flower parts
- tendrils

Some leaves are edible:
- lettuce
- spinach
- parsley

Vocabulary

- ☐ **cuticle**—A waxy covering on the top surface of a leaf; makes the leaf surface somewhat waterproof.
- ☐ **guard cells**—Cells around the stomata that control if it is open or closed.
- ☐ **palisade layer**—The leaf tissue just below the epidermis.
- ☐ **phloem**—Vascular tissue in plants that carries food.
- ☐ **photosynthesis**—The process by which living things use chlorophyll and make their own food (sugar) using the energy of sunlight, water, and carbon dioxide, and give off oxygen (as a waste product).
- ☐ **spongy layer**—The leaf tissue below the palisade layer that has air spaces.
- ☐ **stomata**—Small pores on the underside of a leaf.
- ☐ **xylem**—Vascular tissue that carries water and minerals in plants.

Characteristics

- Leaves are protected by a cuticle over the upper epidermis.
- A vein in a leaf contains xylem and phloem.
  - xylem cells are larger and carry water up from the roots and stem
  - phloem cells are smaller and carry food down to the stem and roots
- Photosynthesis mostly takes place in the palisade layer and the spongy layer.
- The bottom surface of the leaf has tiny openings controlled by guard cells and through which carbon dioxide enters the leaf.

Leaves

LOWER CHALLENGE

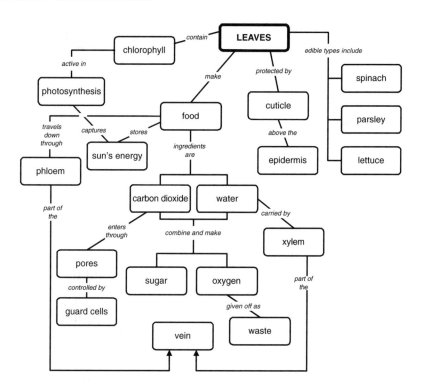

Score: 16 words

Starting hints: The seed item *pores* is where *carbon dioxide* enters the leaf. The connector *make* under the title item *LEAVES* points to *food*.

Notes: The lower-challenge map stresses the role of the leaf in photosynthesis, its materials and products. A few leaf parts are mentioned, but not in detail.

HIGHER CHALLENGE

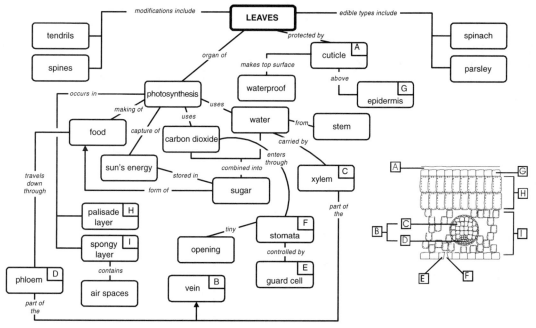

Score: 24 (15 words + 9 letters)

Starting hints: *Water* is carried by the seed item *xylem*. The seed item *opening* references *stomata*.

Notes: The higher-challenge map references a diagram of the cross section of a leaf. You may want to make the diagram-matching portion optional. When checking maps, make sure that the letters match the words in the boxes, not necessarily the position on the map. Students may misplace a word on the map but correctly match the word with the diagram.

Diagram Key
- A cuticle
- B vein
- C xylem
- D phloem
- E guard cell
- F stomata
- G epidermis
- H palisade layer
- I spongy layer

Concept Map: Flowers

Directions: Select words from the word list to fill in the blank map items. Use each word only once, and use all the words on the list.

WORD LIST

anther, birds, filament, fruit, nectar, ovary, petals, pistil, pollen, reproduction, seeds, stigma, style

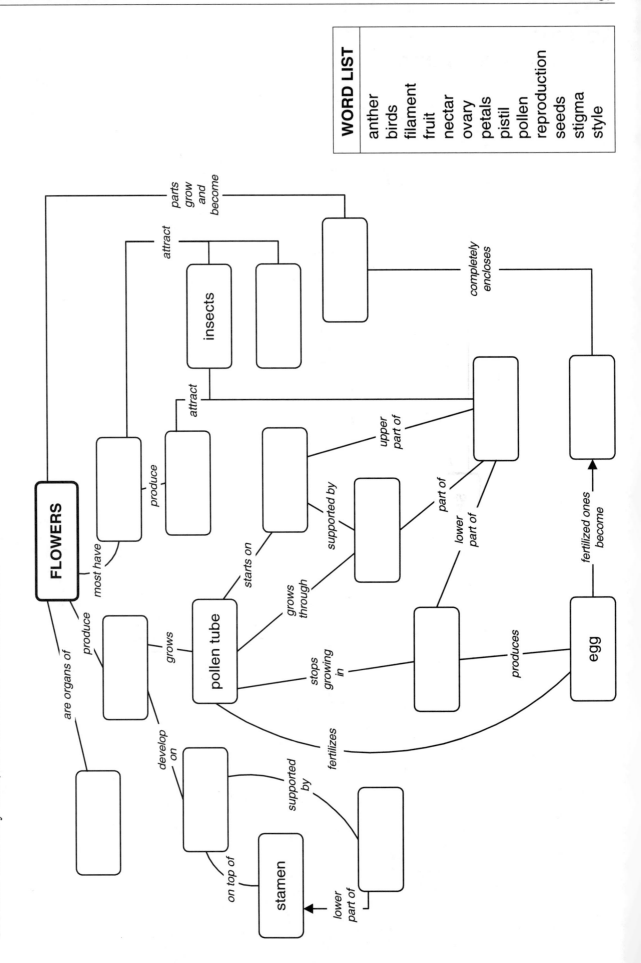

Concept Map: Flowers

THINKING CONNECTIONS: Life Science Book B — **Plant Biology**

Name _____
Date _____ Period _____

Directions: Select words from the word list to fill in the blank map items. Use each word only once, and use all the words on the list. Write the letter of each label on the diagram in the box with its corresponding name.

WORD LIST

anther
egg
female gametophyte
filament
fruit
insects
male gametophyte
nectar
ovary
petals
pistil
pollen
receptacle
reproduction
sepals
style

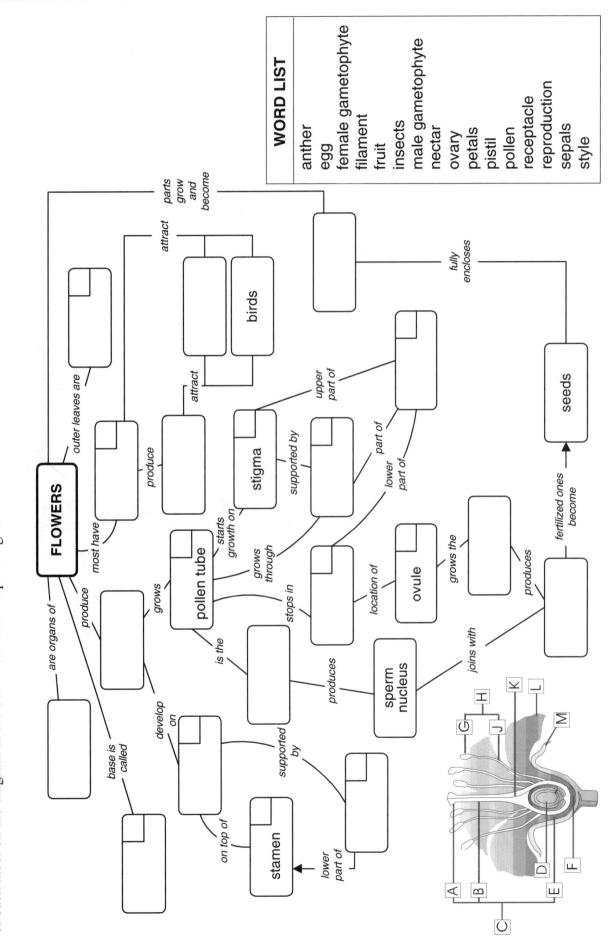

© 1998 Critical Thinking Books & Software • www.criticalthinking.com • (800) 458-4849

81

THINKING CONNECTIONS: Life Science Book B — **Plant Biology**

Critical Thinking →  CONCEPT FILE

Flowers

Background

For many plants, the flower is the organ of reproduction.

All flowering plants produce seeds.

All flowering plants produce fruit that completely cover the seeds.

Vocabulary

- **pistil**—The center of the flower made up of the stigma, the style, and the ovary.
- **pollination**—The transfer of pollen grain from the anther to the stigma.
- **sperm nucleus**—One of several nuclei in the pollen tube; it combines with the egg and fertilizes it.
- **stamen**—The pollen-producing part of the flower, made up of the anther and the filament.

Structure

- All parts of the flower are modified leaves.
- The flower sits on a receptacle.
- The outside leaves are the sepals.
- Petals attract birds and insects by
  – their colors
  – their nectar (sweet fluid)
- The stamen produces pollen in the anther on top of a thin filament.
- The ovary, at the bottom of the pistil and under its style, produces ovules.

Reproduction

- Pollen grains leave the anther and usually travel by air, bird, or insect.
- Pollen grains that land on the top of the stigma (top of the pistil) grow pollen tubes.
- The pollen tube, which is the male gametophyte, stops at the ovule, deep in the ovary.
- The female gametophyte grows in the ovule and produces an egg.
- A sperm nucleus in the pollen tube moves to the egg and fertilizes it.
- Fertilized eggs grow and become seeds.
- Other parts of the flower grow, enclose the seeds, and become the fruit.

Flowers

LOWER CHALLENGE

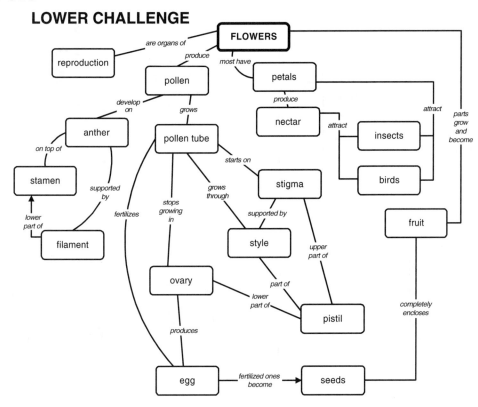

Score: 13 words

Starting hints: The *anther* is *on top of* the seed item *stamen*. The *ovary* produces the seed item *egg*.

Notes: The map may be easier to read if you suggest that students develop a color key and color in the "stamen" branch and the "pistil" branch.

HIGHER CHALLENGE

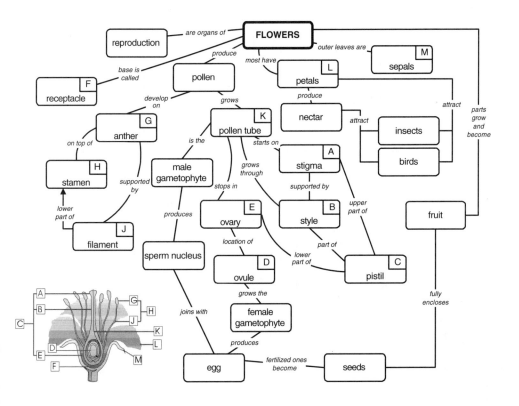

Score: 28 (16 words + 12 letters)

Starting hints: The *anther* is *on top of* the seed item *stamen*. The seed item *stigma* is *supported by* the *style*.

Notes: The higher-challenge map references a flower diagram. Give students time to evolve their own contexts for the information on this busy map. Suggest that students find a scheme for coloring the map and then develop a color key.

You may want to make the diagram-matching portion of the map optional. When checking maps, make sure that the letters match the words in the boxes, not necessarily the position on the map.

Diagram Key

A	stigma
B	style
C	pistil
D	ovule
E	ovary
F	receptacle
G	anther
H	stamen
J	filament
K	pollen tube
L	petal
M	sepal

(Note: The letter *I* was not included to avoid confusion with the letter *L*.)

Concept Map: Sponges

THINKING CONNECTIONS: Life Science Book B
Animal Biology

Name _____
Date _____ Period _____

Directions: Select words from the word list and fill in the blank map items. Use each word only once, and use all the words on the list. Then use two different highlighters, colored pencils, or crayons to color in items that are (1) related to the sponge skeleton and (2) related to sponge reproduction. Show your color scheme in the legend.

COLOR LEGEND

☐ Related to sponge skeletons
☐ Related to sponge reproduction

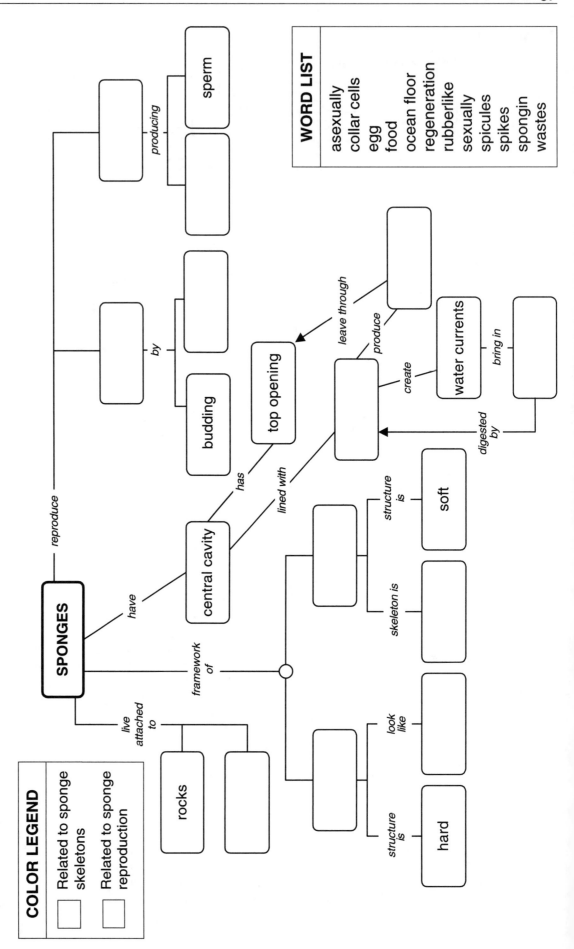

WORD LIST

asexually
collar cells
egg
food
ocean floor
regeneration
rubberlike
sexually
spicules
spikes
spongin
wastes

Concept Map: Sponges

Directions: Select words from the word list and fill in the blank map items. Use each word only once, and use all the words on the list.

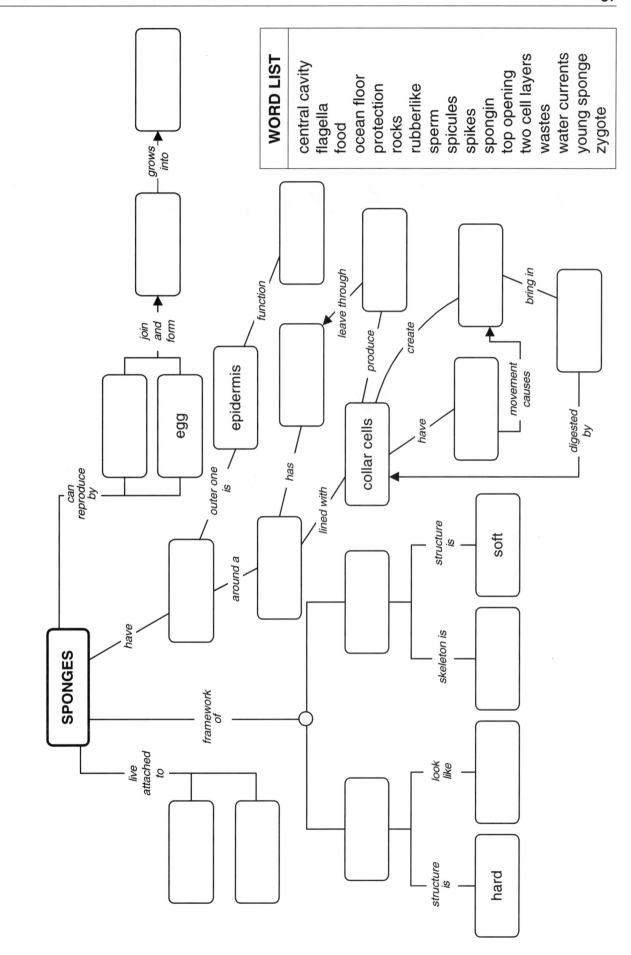

THINKING CONNECTIONS: Life Science Book B — Animal Biology

Critical Thinking →

Sponges

Background

Sponges
- are aquatic animals
- are multicellular
- live attached to ocean floor and to rocks

Digestion

- Digestion occurs in a central cavity.
- Cavity has only one opening, usually on top.
- The cavity is lined with collar cells. These collar cells
  - have flagella, which cause water currents that bring water and food in and out of the cavity
  - do the actual digesting
  - produce wastes, which leave the sponge through the top opening

Reproduction

Sponges reproduce
- sexually
  - sperm and egg produced
  - sperm and egg join and form a zygote
  - zygote grows into young sponge
- asexually
  - by regeneration
  - by budding

Vocabulary

- ❏ **budding**—Reproduction where a young organism grows from the body of its parent.
- ❏ **flagellum** (plural: flagella)—Whiplike structure; can move and create currents; made of fibers of protein.
- ❏ **regeneration**—The growth of a new organism from a section of an existing organism.

Skeletal Framework

Sponges have a skeleton of either
- spongin
  - rubberlike
  - soft
- spicules
  - hard
  - spikelike in appearance

The outer surface of sponge is the epidermis, which protects the sponge.

Sponges

LOWER CHALLENGE

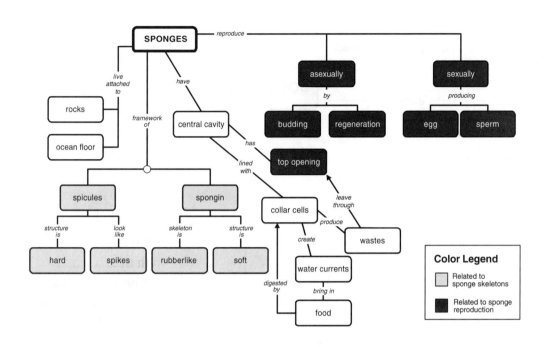

Score: 25 (12 words + 13 colored items)

Starting hints: A good starting place here is the connector label *digested by*.

Notes: The lower challenge map explores sponge reproduction.

Remember that the map notation O means "or."

HIGHER CHALLENGE

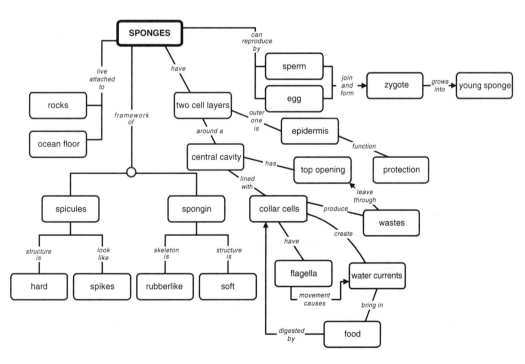

Score: 17 words

Starting hints: The seed item *collar cells* line the *central cavity*. The seed item *egg* joins with *sperm*, forming the *fertilized egg*.

Notes: The items *ocean floor* and *rocks* are interchangeable.

THINKING CONNECTIONS: Life Science Book B — Animal Biology

Concept Map: Coelenterates

Name _____

Date _____ Period ____

Directions: Select words from the word list and fill in the blank map items. Cut and paste (or tape) the pictures in the correct boxes on the map. Use each word and picture only once, and use all the words on the list. Be sure to place words relating to reproduction and food in the correctly shaped boxes.

COELENTERATES
- examples
- reproduction
- body contains → digestive cavity
  - has → (box)
    - is a → (box) surrounded by → tentacles (contain → box → paralyze)
    - is an → (box) eliminates → wastes
  - takes in → (box) ← becomes

examples:
- (box) lives in → (box)
- (box) movement → free swimming
- (box) builds → reefs

reproduction:
- methods → often reproduces by → regeneration → growth of → (box)
- production of → egg

PICTURE GALLERY
- jellyfish
- coral
- Hydra

WORD LIST
anus	mouth
asexual	prey
budding	single opening
food	sexual
fresh water	sperm
lost body parts	stinging cells

SHAPE LEGEND
- ▽ Relating to reproduction
- ▭ Relating to food

THINKING CONNECTIONS: Life Science Book B — Animal Biology

Concept Map: Coelenterates

Name _____

Date _____ Period ___

Directions: Select words from the word list and fill in the blank map items. Cut and paste (or tape) the pictures in the correct boxes on the map. Use each word and picture only once, and use all the words on the list. Then use two different highlighters, colored pencils, or crayons to color in items that are (1) related to food and (2) related to reproduction. Show your color scheme in the legend.

PICTURE GALLERY
- jellyfish
- coral
- Hydra

COLOR LEGEND
- ☐ Related to food
- ☐ Related to reproduction

WORD LIST

anus	free swimming	reefs
asexual	fresh water	regeneration
budding	lost body parts	sexual
digestive cavity	mouth	sperm
ectoderm	prey	tentacles
endoderm	protection	wastes
food		

© 1998 Critical Thinking Books & Software • www.criticalthinking.com • (800) 458-4849

89

THINKING CONNECTIONS: Life Science Book B   Animal Biology

Critical Thinking → CONCEPT FILE

Coelenterates

Examples

Jellyfish
- free-swimming in the ocean
- clear and jellylike

Hydra
- attaches to bottom or leaves but can somersault
- reproduces by budding
- lives in fresh water

Coral
- is tiny
- lives in colonies

The dead bodies under the living bodies form reefs.

Vocabulary

- ❐ **anus**—An opening through which solid wastes leave the body.
- ❐ **budding**—Reproduction where a young organism grows from the body of its parent.
- ❐ **regeneration**—Growth of lost body parts.
- ❐ **stinging cells**—Cells that produce poison that can paralyze other animals.
- ❐ **tentacles**—Structures around the opening of the digestive system that contain stinging cells.

Characteristics

Coelenterates
- have two tissue layers
  – an inner endoderm, lining the digestive cavity
  – an outer ectoderm, providing protection
- reproduce
  – sexually, producing sperm and egg
  – asexually, by budding and by regeneration
- digest food in a digestive cavity
  – One opening at the top does the job of both mouth and anus.
  – The mouth is usually surrounded by tentacles.

Coelenterates

LOWER CHALLENGE

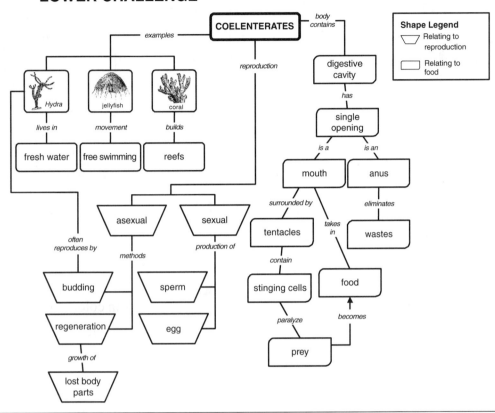

Score: 15 (12 words + 3 pictures)

Starting hints: The *anus eliminates* the seed item *wastes*. The seed item *tentacles* contains the *stinging cells* that *paralyze* the item *prey*. Of the three pictures, it's the jellyfish that has *free swimming* movement.

Notes: The Shape Legend is provided here as another enabler for the students.

HIGHER CHALLENGE

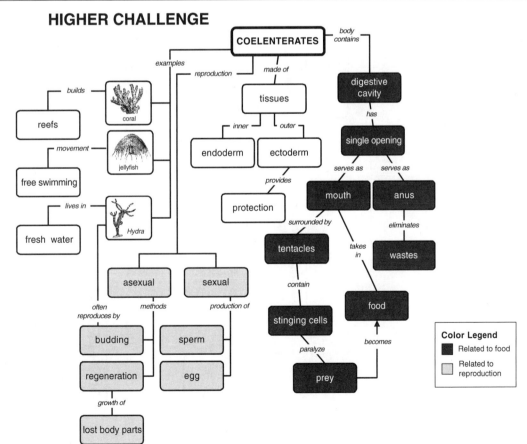

Score: 38 (19 words + 3 pictures + 16 colored items)

Starting hints: The seed item *stinging cells* are contained on the *tentacles* and *paralyze prey*. Paired above the seed item *egg* would be *sperm*, both of which point to *sexual* reproduction.

Concept Map: Worms

Directions: Select words from the word list and fill in the blank map items. Use each word only once, and use all the words on the list. Then use two different highlighters, colored pencils, or crayons to color in items that are (1) related to flatworms, (2) related to roundworms, and (3) related to segmented worms. Show your color scheme in the legend.

WORD LIST

anus	food	pharynx
blood	food grinding	*Planaria*
digestion	gizzard	roundworms
earthworm	intestines	segmented worm
flatworm	mouth	

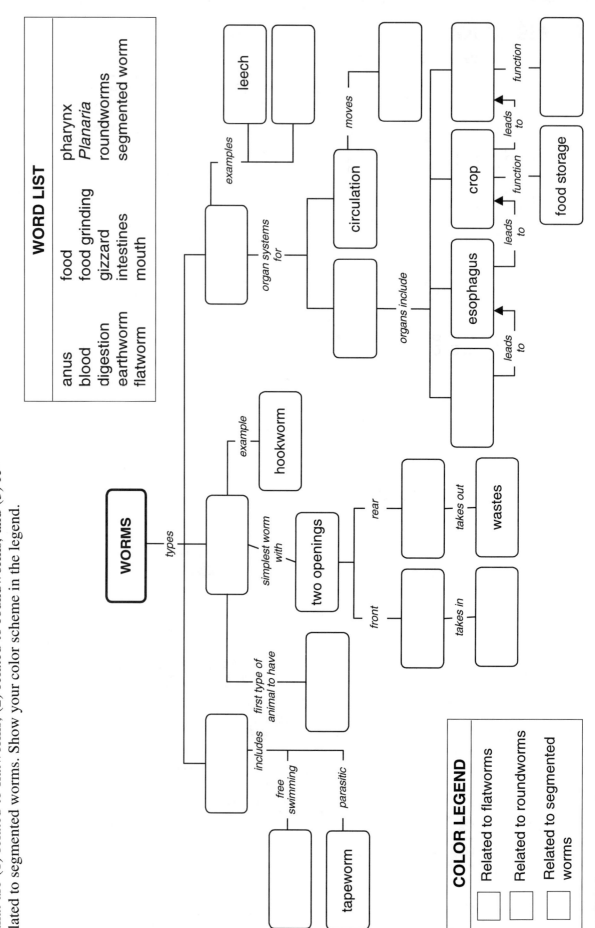

COLOR LEGEND

- Related to flatworms
- Related to roundworms
- Related to segmented worms

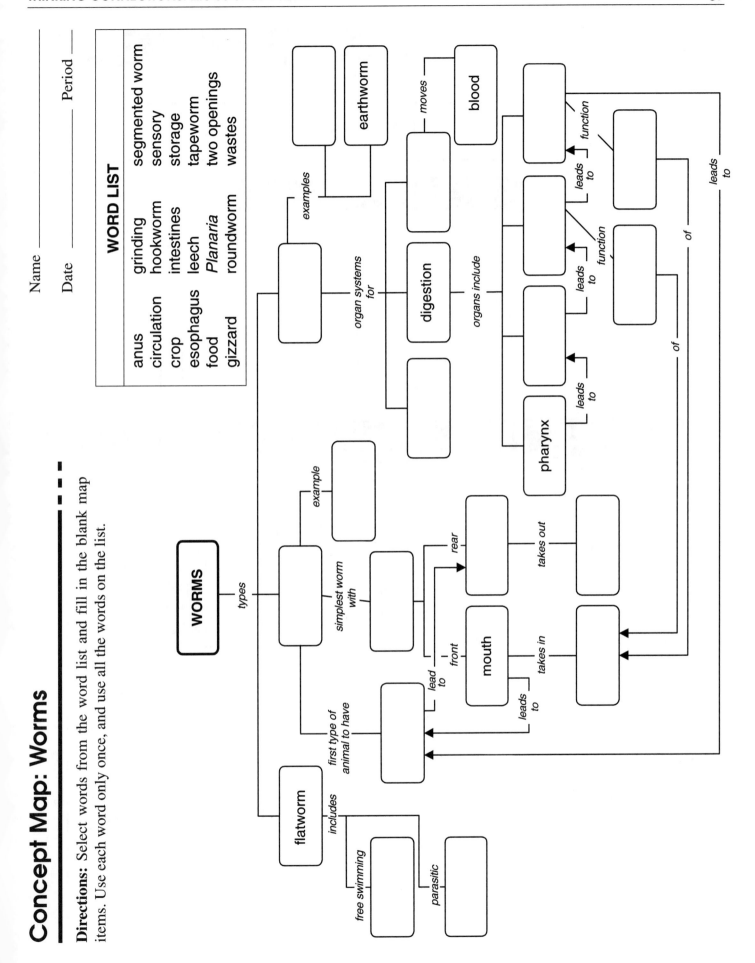

Critical Thinking → Concept File

Worms

Vocabulary

Check your understanding—these terms are explained on this page.

- ❏ anus
- ❏ crop
- ❏ gizzard
- ❏ hookworm
- ❏ intestine
- ❏ leech
- ❏ *Planaria*
- ❏ tapeworm

Types

Segmented worms

- have major organ systems for
  - digestion
  - senses
  - circulation of blood

Their organs of digestion include
- pharynx
- esophagus
- crop, where food is stored
- gizzard, where food is ground up
- intestine, where food is absorbed

Examples
- earthworm
- leech

Flatworms

- include
  - *Planaria*, a free-swimming worm
  - tapeworm, a parasite

Roundworms

- are the simplest worms with two openings in the digestive system
  - mouth, the front opening through which food enters the animal
  - anus, the rear opening through which wastes are eliminated
- are the first type of animal to have an intestine

An example is the hookworm.

Worms

LOWER CHALLENGE

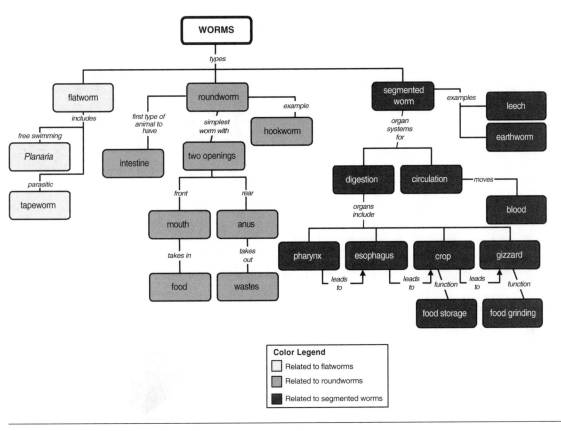

Score: 37 (14 words + 23 colored items)

Starting hints: The seed item *hookworm* is an *example* of *roundworms*. The seed item *food storage* is a *function* of the *gizzard*.

Notes: The color scheme will help students read the map. Because the color scheme follows the map categories directly, however, you may not want to count it as part of the score.

HIGHER CHALLENGE

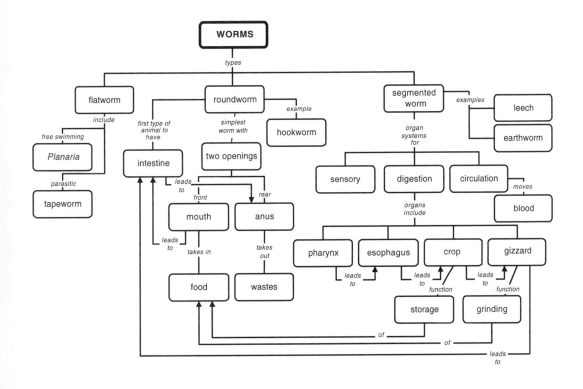

Score: 18 words

Starting hints: The seed item *blood* is moved by the *circulation* system. The seed item *mouth* is one of *two openings*; the other is the *anus*.

Notes: The many cross connections on this map make it harder to read at first, so give students plenty of time to develop a context for the map items. You might encourage students to color the three main branches of this map (as above). It will make the map much easier to read and review.

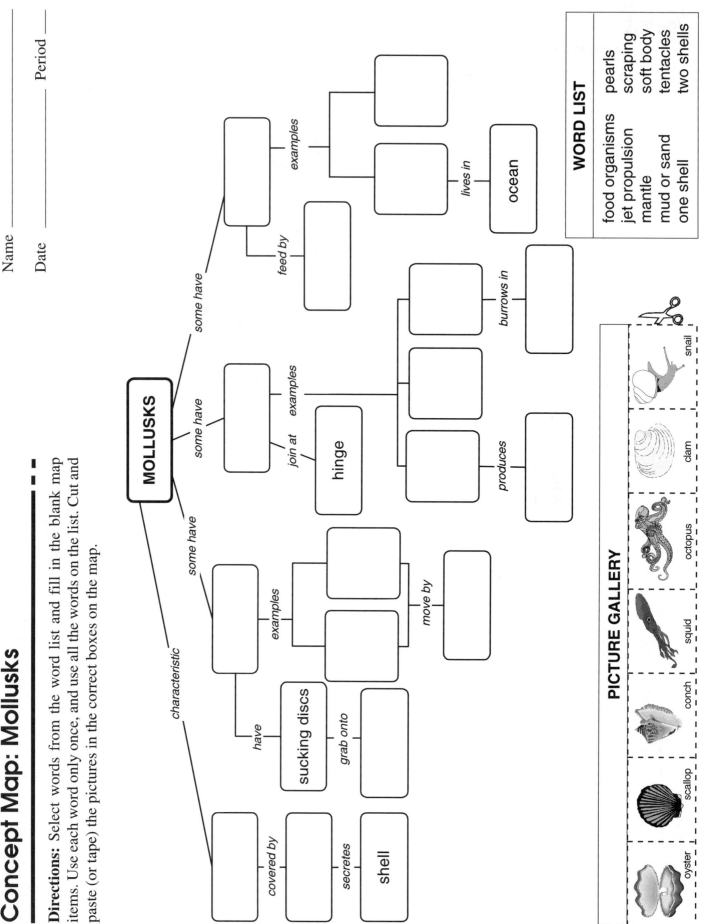

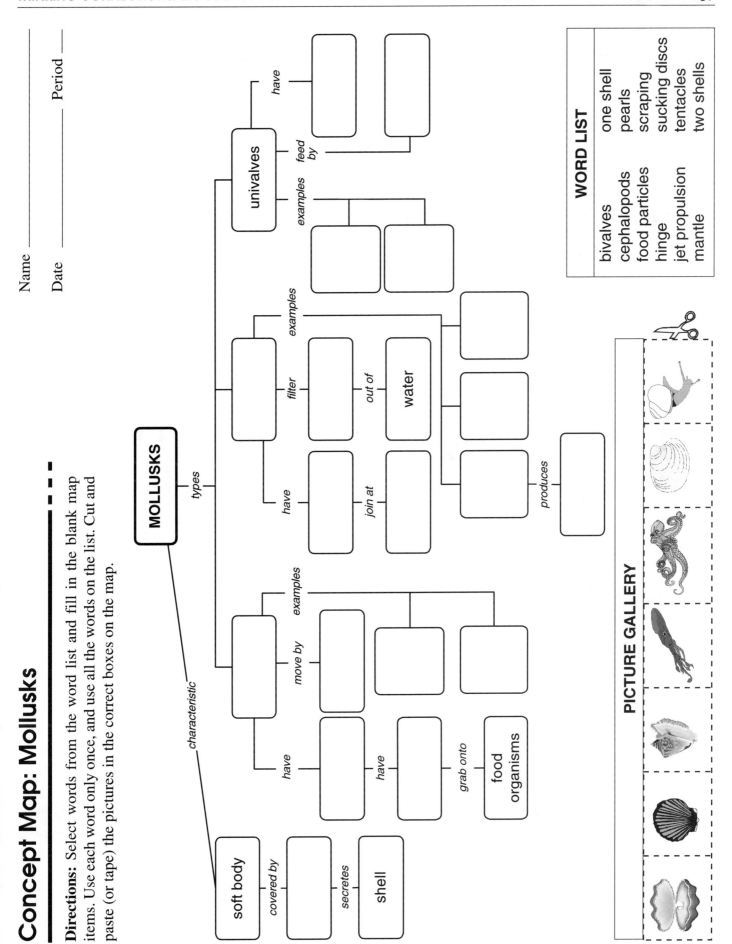

THINKING CONNECTIONS: Life Science Book B — Animal Biology

Critical Thinking →  CONCEPT FILE

Mollusks

Characteristics

- soft body covered by a mantle, an outside covering that in many cases produces a shell
- many forms have shells
  - univalves have one shell
  - bivalves have two shells

Vocabulary

Check your understanding—these terms are explained on this page.

- ❒ bivalve
- ❒ mantle
- ❒ sucking disc
- ❒ tentacle
- ❒ univalve

Groups

Univalves

- They feed by scraping food off surfaces.
- The snail and the ocean-living conch are examples.
- A slug is a univalve without a visible shell.

Cephalopods

- The mouth has tentacles around it.
- The tentacles have sucking discs that grab and hold food organisms.
- Many swim by jet propulsion.
- Squid and octopus are examples.

Bivalves

- The shells are joined at a hinge.
- They feed by filtering water and extracting food particles.
- The oyster, clam, and scallop are examples.
- Pearls are produced by a bivalve (oyster).
- Some burrow in the sand (for example, clam).

© 1998 Critical Thinking Books & Software • www.criticalthinking.com • (800) 458-4849

Mollusks

LOWER CHALLENGE

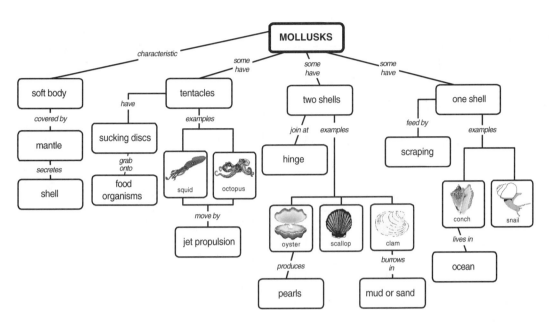

Score: 17 (10 words + 7 pictures)

Starting hints: The connector *move by* leads to *jet propulsion*. The connector *burrows in* leads to *mud or sand*.

Notes: Although there is not a Color Legend on this map, students may find it easier to read and review if they color each branch and make their own legend.

The squid and the octopus are interchangeable by position.

HIGHER CHALLENGE

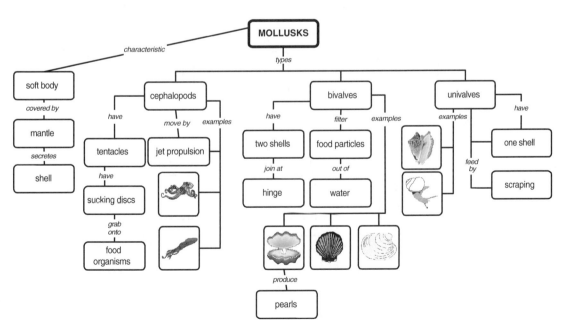

Score: 19 (12 words + 7 pictures)

Starting hints: The seed item *shell* is secreted by the *mantle*. The connector *grab onto* strongly suggests *tentacles*.

Notes: This version uses class names as an organizing theme. Also, the pictures do not carry with them identifying labels.

The scallop and the clam are interchangeable by position, as are the octopus and the squid, and the conch and the snail.

Concept Map: Arthropods

Name _____
Date _____ Period _____

Directions: Select words from the word list and fill in the blank map items. Use each word only once, and use all the words on the list. Then use two different highlighters, colored pencils, or crayons to color in items that are (1) related to only arachnids and (2) related to only crustaceans. Show your color scheme in the legend.

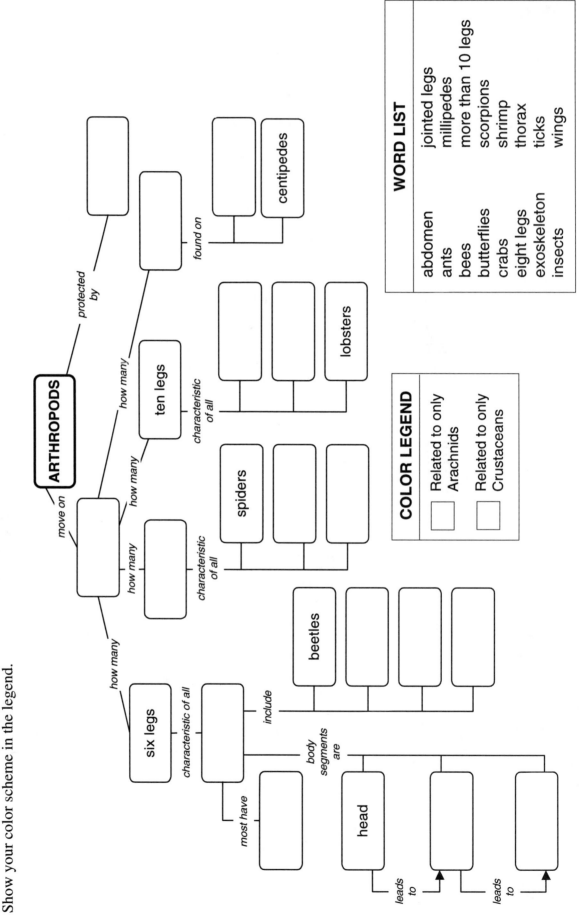

Concept Map: Arthropods

Directions: Select words from the word list and fill in the blank map items. Use each word only once, and use all the words on the list.

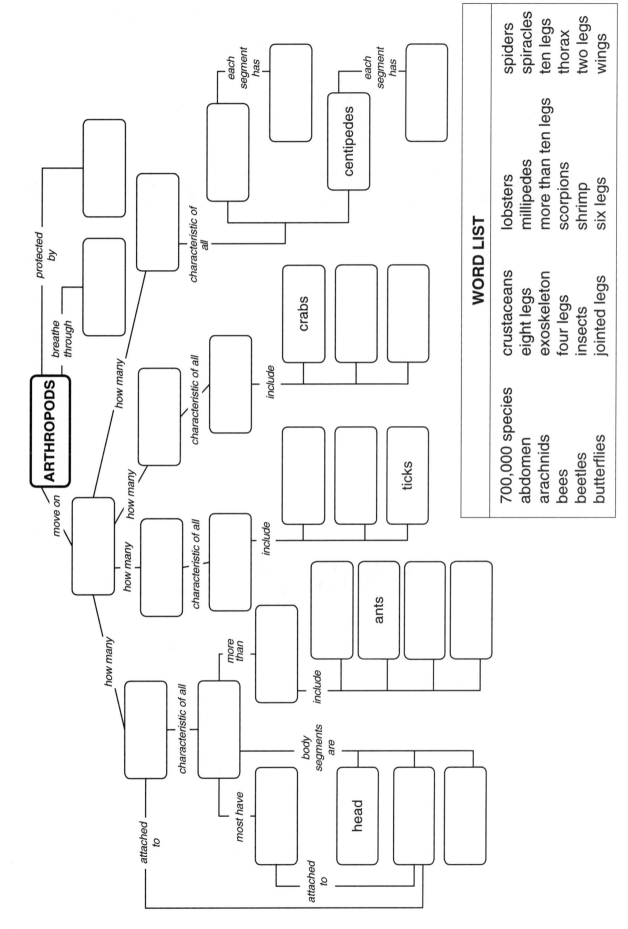

THINKING CONNECTIONS: Life Science Book B — Animal Biology

Critical Thinking →  CONCEPT FILE

Arthropods

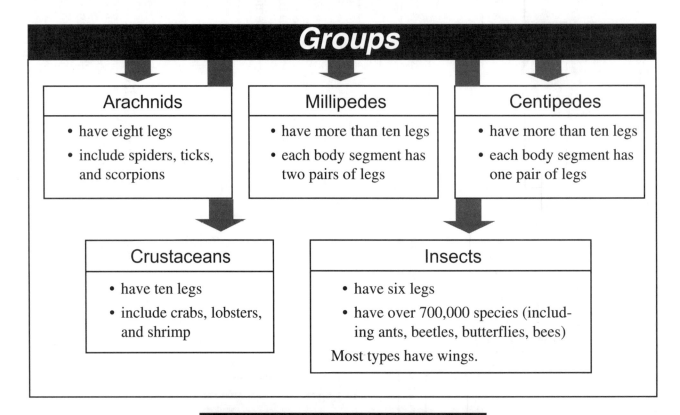

Vocabulary

- **exoskeleton**—A skeleton on the outside of the body.
- **abdomen**—The back section of an arthropod's body.
- **thorax**—The middle section of an insect's body to which legs and wings are attached.
- **spiracles**—Holes in the bodies of many arthropods through which they breathe.

Arthropods

LOWER CHALLENGE

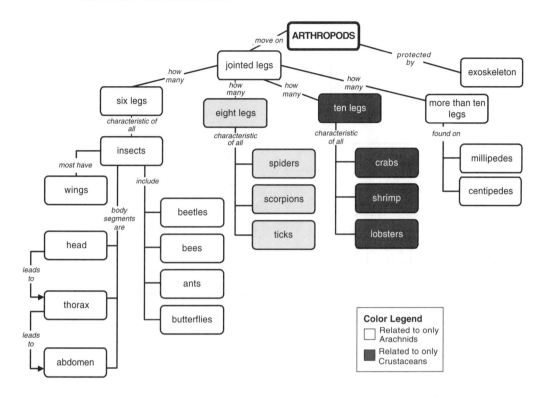

Score: 24 (16 words + 8 colored items)

Starting hints: The seed item *six legs* points to *insects*. The seed item *ten legs* points to *shrimp* and *crabs*.

Notes: This map uses number of legs to distinguish the various arthropod groups. Some of the specific class names are given in the Color Legend.

The Color Legend uses the word *only* to limit what can be included. For example, Arachnids have thoraxes, but so do other arthropods.

In most cases, names of types of animals within groups can be placed in any order.

HIGHER CHALLENGE

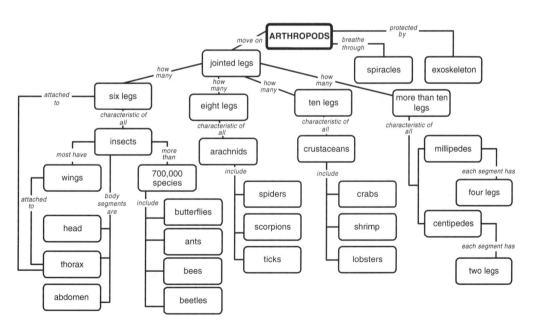

Score: 24 words

Starting hints: The seed item *ticks* points to *arachnids*. The seed item *crabs* points to *crustaceans*. The connector *breathe through* points to *spiracles*.

Notes: Only five items are given at the start of this map, so students will have to use lots of deductive reasoning to get the items placed correctly.

For the most part, names of types of animals within groups can be in any order.

Concept Map: Cold-blooded Vertebrates

Directions: Select words from the word list and fill in the blank map items. Use each word only once, and use all the words on the list.

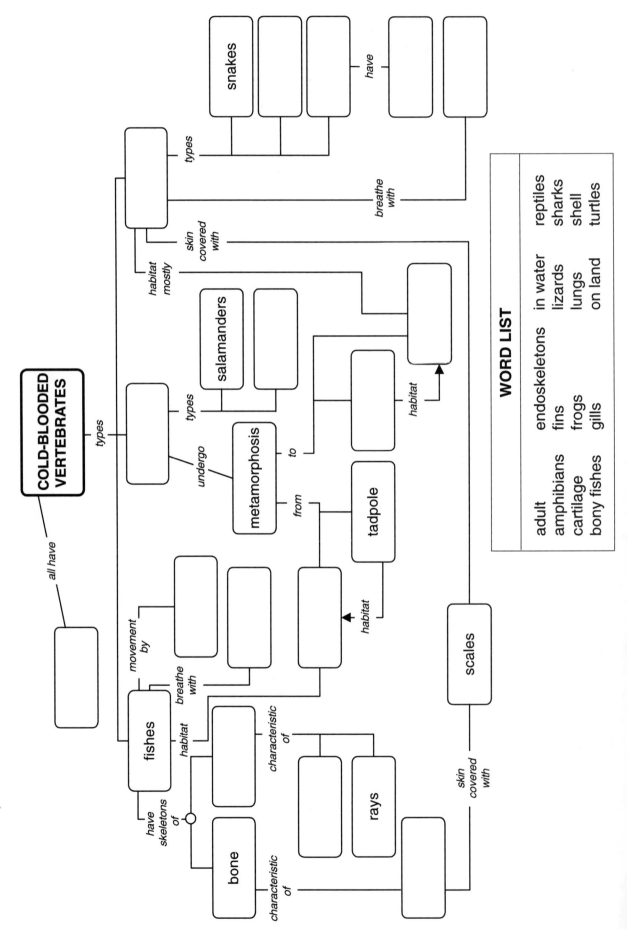

WORD LIST

adult	endoskeletons	in water	reptiles
amphibians	fins	lizards	sharks
cartilage	frogs	lungs	shell
bony fishes	gills	on land	turtles

Concept Map: Cold-blooded Vertebrates

Directions: Select words from the word list and fill in the blank map items. Use each word only once, and use all the words on the list.

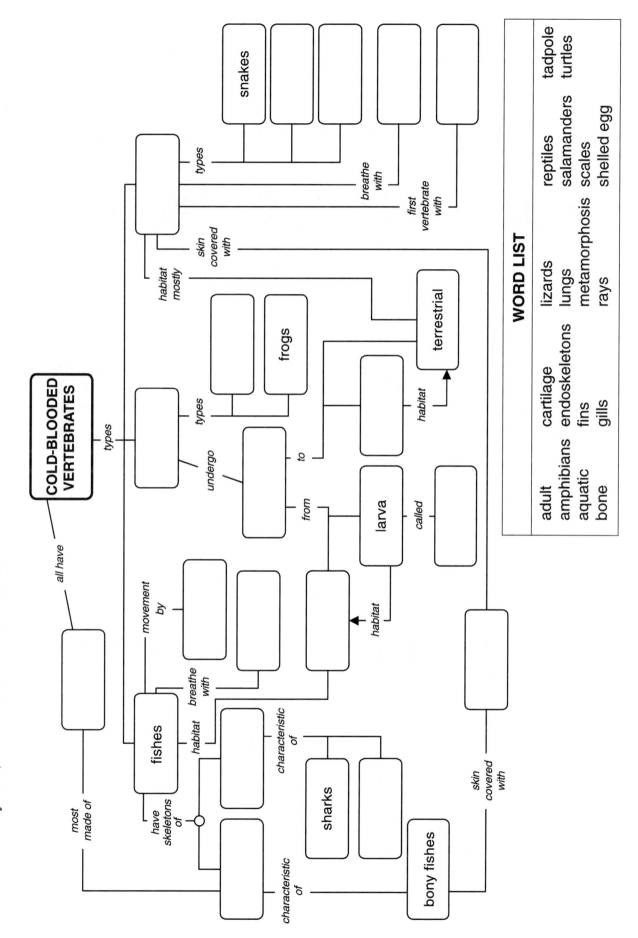

Critical Thinking

Cold-blooded Vertebrates

Major Types

Fish
- All forms are aquatic.
- Bony fishes have skeletons of bone.
  – Most have scaly skin.
- Other fishes have skeletons of cartilage, including
  – sharks
  – rays
- Fish breathe through gills.
- They move with fins.

Amphibians
- are both aquatic and terrestrial
- undergo metamorphosis
  – The immature larva stage (usually aquatic) grows into the adult stage (usually terrestrial).
- include
  – frogs
  – toads
  – salamanders

Reptiles
- have scaly skin
- were first type of vertebrate to develop an egg with a shell
- live on land
- breathe with lungs
- include
  – snakes
  – lizards
  – turtles

Vocabulary

- ❏ **aquatic**—Living in water.
- ❏ **cartilage**—A thick, tough, flexible tissue softer than bone.
- ❏ **endoskeleton**—A skeleton on the inside of an animal's body, found in all vertebrates.
- ❏ **metamorphosis**—The process in which an animal loses one body form and develops another.
- ❏ **tadpole**—The larva stage of amphibians.
- ❏ **terrestrial**—Living on land.

Cold-blooded Vertebrates
LOWER CHALLENGE

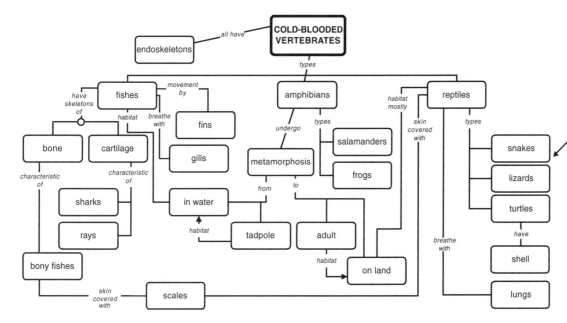

Score: 16 words

Starting hints: The seed item *salamanders* is a type of *amphibian*. The seed item *fishes* leads into *breathe with gills*. The habitat of the *tadpole* is *in water*.

Notes: The examples within each group are interchangeable except for the reptile group, which must be as shown.

If students add color and a Color Legend to the map, it may improve readability. There are, however, a number of items shared by major branches, so a possible Color Legend scheme might be as follows:

Color Legend
☐ Related to fishes
☐ Related to amphibians
☐ Related to reptiles
☐ Related to fishes and reptiles
☐ Related to fishes and amphibians
☐ Related to amphibians and reptiles

HIGHER CHALLENGE

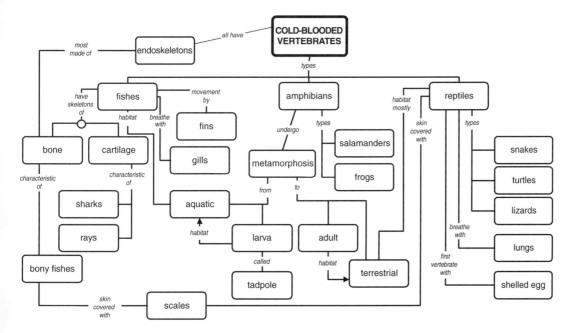

Score: 18 words

Starting hints: The seed item *frogs* is a type of *amphibian*. The seed item *bony fishes* have skin covered with *scales*. The seed item *fishes* links to *breathe with gills*.

Notes: If students add color and a Color Legend to the map, it may improve readability. There are, however, a number of items shared by major branches, so a possible Color Legend scheme might look like the above.

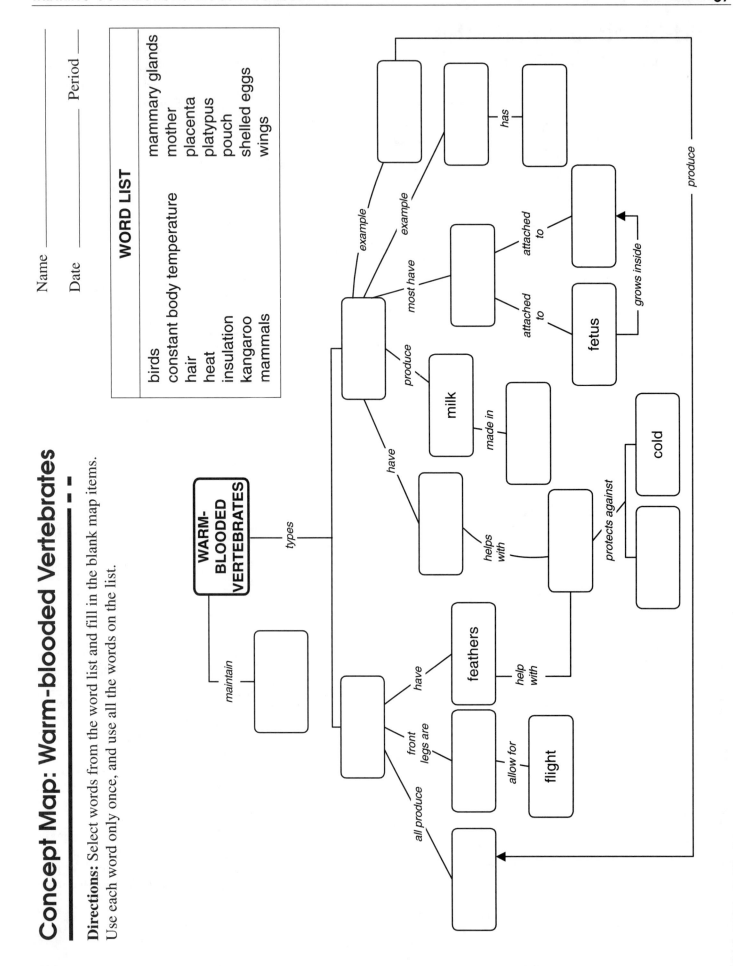

Concept Map: Warm-blooded Vertebrates

Directions: Select words from the word list and fill in the blank map items. Use each word only once, and use all the words on the list.

WORD LIST

birds	hair
constant body temperature	heat
feathers	insulation
fetus	kangaroo
flight	mammals
	marsupials
	milk
	monotreme
	mother
	placenta
	platypus
	shelled egg
	wings

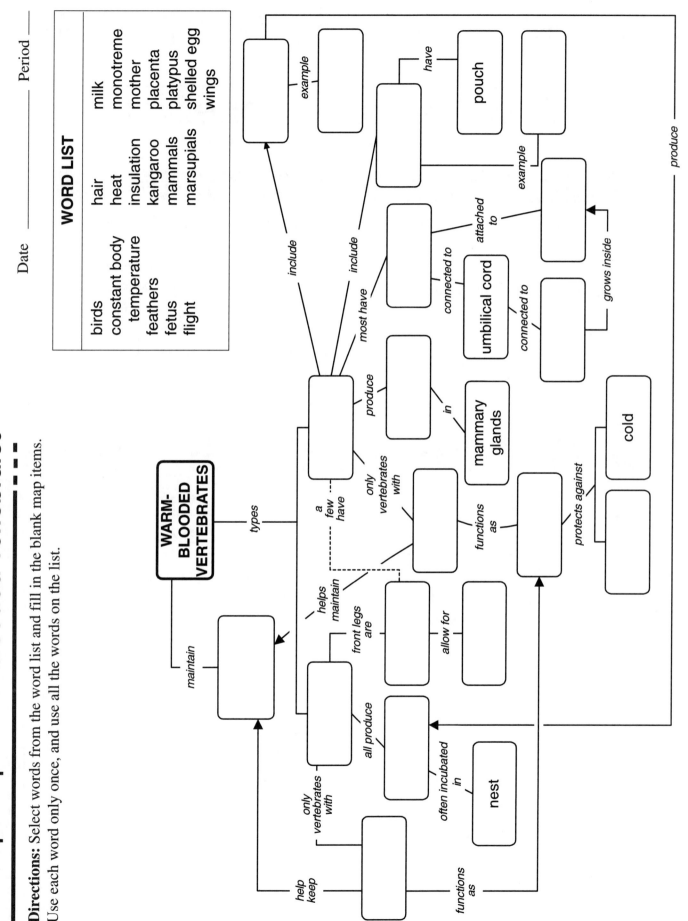

Critical Thinking

Warm-blooded Vertebrates

Characteristics

- have a backbone
- maintain a constant body temperature
- are insulated by body coverings
  – feathers
  – hair

Vocabulary

- ❏ **insulation**—Material that slows or stops the loss or gain of heat.
- ❏ **mammary glands**—Glands in the female mammal that produce milk.
- ❏ **monotremes**—A group of mammals that produce shelled eggs.
- ❏ **placenta**—A structure that attaches (by the umbilical cord) the fetus to the mother in most mammals.

Types

Mammals

- produce milk in mammary glands
- are the only vertebrates with hair
- are insulated from heat and cold by hair on their bodies

Examples include
- placental mammals
  – all have a placenta
- monotremes
  – example is the platypus
  – all have a shelled egg
- marsupials
  – example is the kangaroo
  – all have a pouch

In most cases, the front legs are not wings (exception, bats).

Birds

- All produce a shelled egg.
- Eggs are often incubated in a nest.
- Front legs are modified into wings, allowing flight.
- These are the only vertebrates with feathers.
- Feathers insulate their bodies against
  – heat
  – cold

Warm-blooded Vertebrates
LOWER CHALLENGE

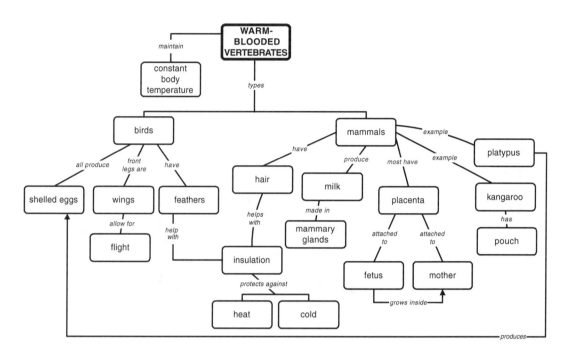

Score: 14 words

Starting hints: Good starting points are the seed items *feathers* and *milk*, which establish the identity for the two major branches on this map.

Notes: This version of the map focuses on the basic characteristics of birds and mammals and avoids order names. The connections are much simpler than on the higher-challenge map.

HIGHER CHALLENGE

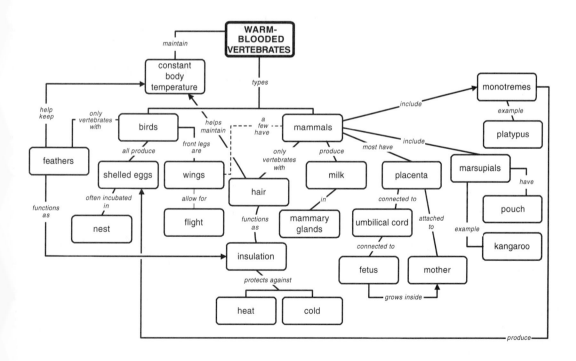

Score: 18 words

Starting hints: The seed items *mammary glands* and *nest* are good starting points. The connector *front legs are* is a strong hint for the "birds" branch.

Notes: This map is rich in interconnections, recognizing that a few mammals have wings and that both feathers and hair help in homeostasis.

Students could improve the readability of this very busy map by adding color and a Color Legend.

Concept Map: Skeletal System

Directions: Select words from the word list and fill in the blank map items. Use each word only once, and use all the words on the list.

WORD LIST

ball-and-socket	joints
body	knee
body organs	marrow
bones	muscles
hinge	red blood cells
hip	skull
immovable	wrist

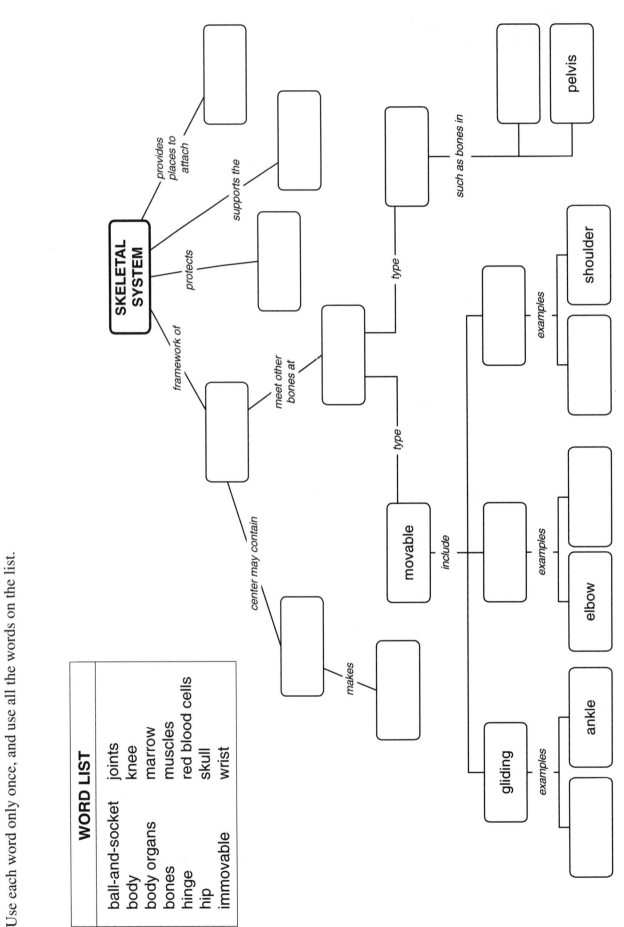

THINKING CONNECTIONS: Life Science Book B — Human Biology

Concept Map: Skeletal System

Directions: Select words from the word list and fill in the blank map items. Use each word only once, and use all the words on the list.

WORD LIST

ball and socket	pelvis
body	pivot
bones	red blood cells
calcium	shoulder
cartilage	skull
elbow	wrist
gliding	
hinge	
immovable	
joints	
marrow	
movable	
muscles	
organs	

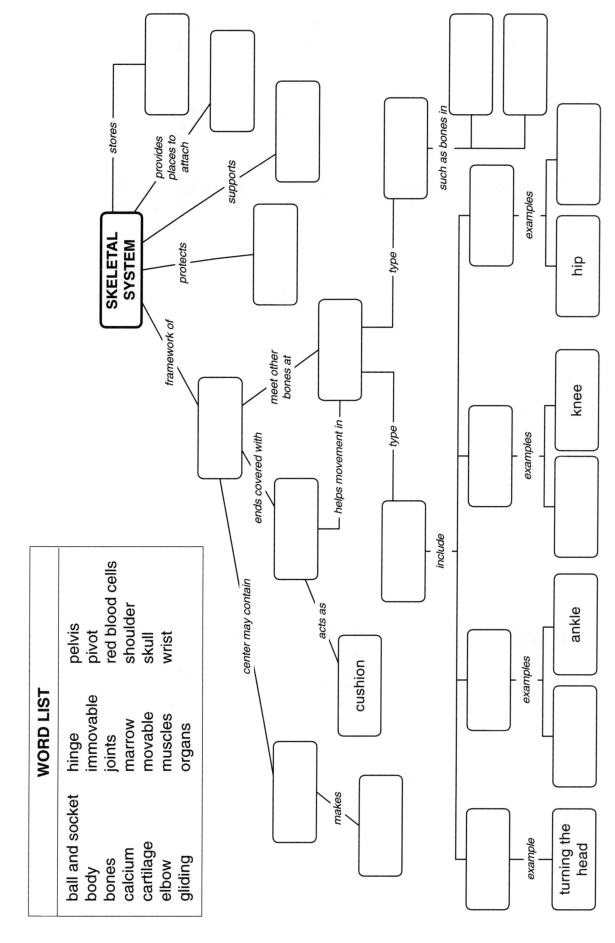

© 1998 Critical Thinking Books & Software • www.criticalthinking.com • (800) 458-4849   113

Critical Thinking → CONCEPT FILE

Skeletal System

Characteristics

- Bones protect internal body organs and tissues.
- Most muscles are attached to bones.
- Bones provide support for the body.
- Bones store calcium.
- The marrow at the center of larger bones makes red blood cells.
- Cartilage covers the ends of bones, cushioning the bones and helping joints to move.
- Joints are formed where bones meet bones; joints are
  - immovable, such as joints in the pelvis and in the skull
  - movable (see below)

Vocabulary

- ❑ **cartilage**—A thick, tough, flexible tissue softer than bone.
- ❑ **immovable joints**—Joints where the bones themselves do not move.
- ❑ **movable joints**—Joints where the bones move by turning, gliding, pivoting, or swinging.

Types of Movable Joints

Hinge	Ball-and-socket	Gliding	Pivot
Examples • knee • elbow	Examples • shoulder • hip	Examples • ankle • wrist	Example • turning the head

Skeletal System

LOWER CHALLENGE

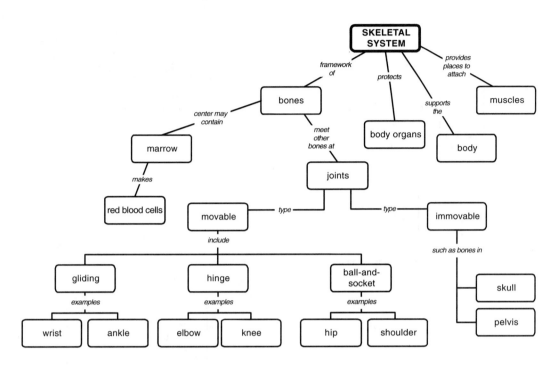

Score: 14 words

Starting hints: The seed item *movable* is a type of *joint*. The seed item *shoulder* is a *ball-and-socket* joint, as is its pair, the *hip*. The two main branches in this map are the "movable" and the "immovable" joints. If students have trouble getting started, you might turn their attention to these terms.

HIGHER CHALLENGE

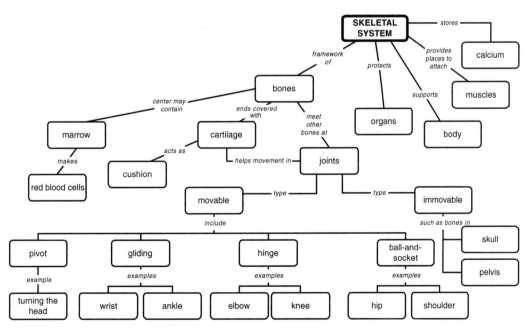

Score: 20 words

Starting hints: The connector *meet other bones at* points to *joints*, which is a major gateway to the bottom branch of this map.

Notes: The higher-challenge map includes six terms not on the lower-challenge map.

Concept Map: Muscular System

Directions: Select words from the word list and fill in the blank map items. Use each word only once, and use all the words on the list.

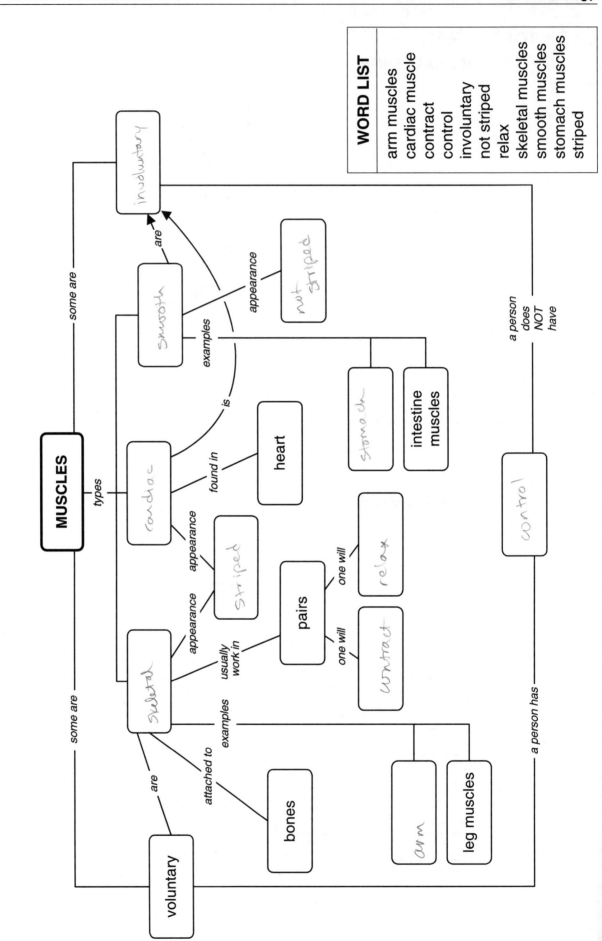

Concept Map: Muscular System

Directions: Select words from the word list and fill in the blank map items. Use each word only once, and use all the words on the list.

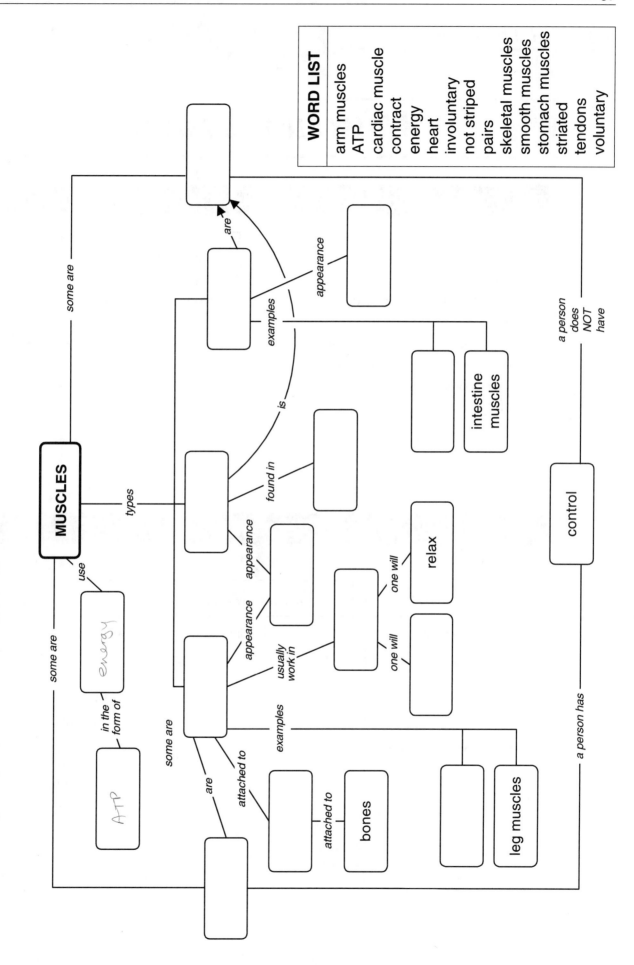

THINKING CONNECTIONS: Life Science Book B — Human Biology

Critical Thinking

Muscular System

Background

- There are three types of muscles in human beings: cardiac, skeletal, and smooth.
- Muscles are either voluntary or involuntary.
- The appearance of muscles is either striped (striated) or non-striped.
- Energy is stored in the muscle cells in the form of ATP.

Vocabulary

- ❏ **involuntary muscle**—A muscle that a person *cannot* consciously control.
- ❏ **voluntary muscle**—A muscle that a person *can* consciously control.

Types of Muscle

Smooth Muscles

- are not striped in appearance
- are involuntary

Examples
- muscles of the stomach
- muscles of the intestines

Cardiac Muscle

- is striped in appearance
- is involuntary
- is found in the heart

Skeletal Muscles

- are attached to bones (tendons connect these muscles to bones)
- are voluntary
- are striated
- work in pairs—bones move when one of the pair contracts and the other relaxes

Examples
- muscles of the arm
- muscles of the leg

118 © 1998 Critical Thinking Books & Software • www.criticalthinking.com • (800) 458-4849

Muscular System

LOWER CHALLENGE

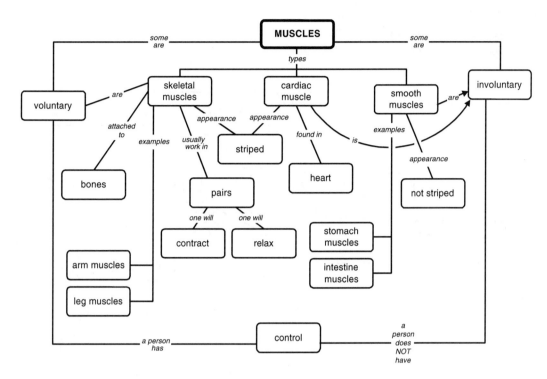

Score: 11 words

Starting hints: Good starting points for this map are the items *heart* and *bones*.

Notes: The items *contract* and *relax* are interchangeable.

HIGHER CHALLENGE

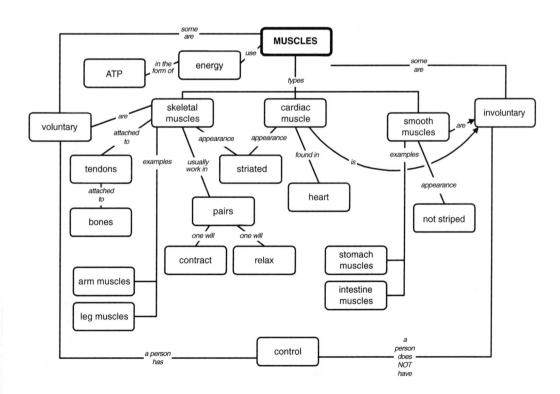

Score: 15 words

Starting hints: Good starting points on this map are the seed items *bones, leg muscles*, and *intestine muscles*.

Notice also the connectors *is* and *are*, which indicate singular and plural items.

THINKING CONNECTIONS: Life Science Book B — Human Biology

Concept Map: Digestive System

Name _____

Date _____ Period ____

Directions: Select words from the word list and fill in the blank map items. Use each word only once, and use all the words on the list. Write the letter of each label on the diagram in the box with its corresponding name.

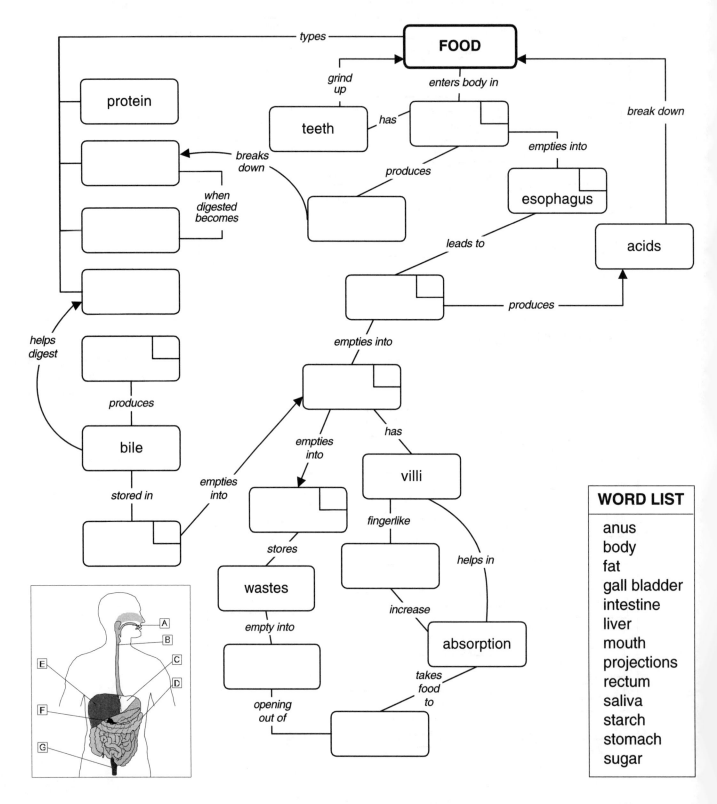

WORD LIST

anus
body
fat
gall bladder
intestine
liver
mouth
projections
rectum
saliva
starch
stomach
sugar

THINKING CONNECTIONS: Life Science Book B — Human Biology

Concept Map: Digestive System

Name _____

Date _____ Period ____

Directions: Select words from the word list and fill in the blank map items. Use each word only once, and use all the words on the list. Write the letter of each label on the diagram in the box with its corresponding name.

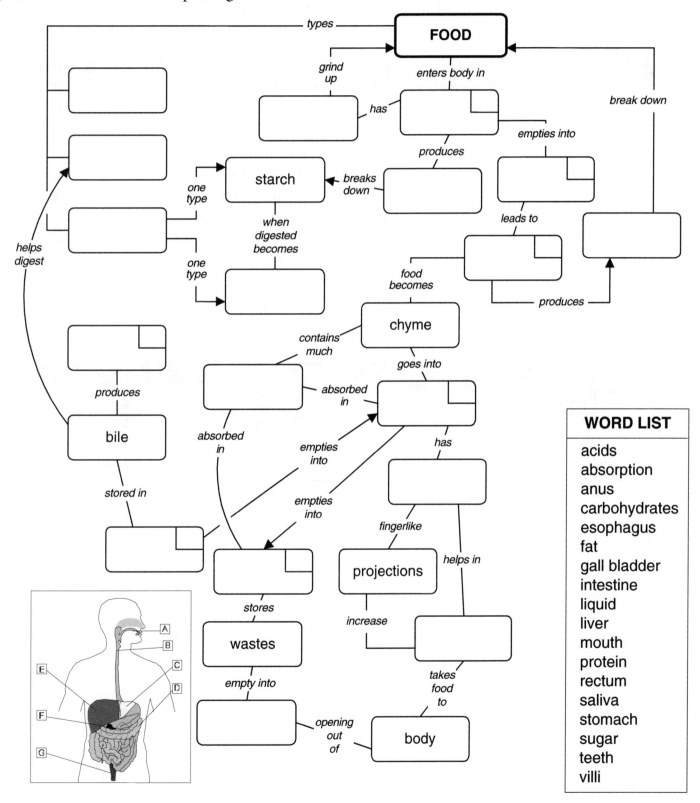

WORD LIST

acids
absorption
anus
carbohydrates
esophagus
fat
gall bladder
intestine
liquid
liver
mouth
protein
rectum
saliva
stomach
sugar
teeth
villi

© 1998 Critical Thinking Books & Software • www.criticalthinking.com • (800) 458-4849

THINKING CONNECTIONS: Life Science Book B — Human Biology

Critical Thinking →  CONCEPT FILE

Digestive System

Foods

There are three major types of foods:
- carbohydrates, which include
  - sugars
  - starches
- fats
- proteins

Vocabulary

- ❑ **anus**—An opening through which solid wastes leave the body.
- ❑ **esophagus**—The tube that connects the mouth to the stomach.
- ❑ **rectum**—A place for solid waste storage and water absorption.
- ❑ **villi** (singular: villus) Fingerlike projections on the inner surface of the intestines.

Major Organs of Digestion

 **Mouth**
- The mouth is the first place of digestion.
- Saliva digests starches into simple sugars.
- Teeth grind up food, allowing easier digestion in the stomach.
- The mouth leads to the esophagus.

 Stomach
- produces acids
- turns food into liquid called "chyme"
- leads to the intestines

 Intestines
- are about 30 feet long
- have villi that increase surface area
- absorb food into the blood
- lead to the rectum

 Liver
- produce bile
- helps to digest fat in the intestines
- stores bile in the gall bladder

Digestive System

LOWER CHALLENGE

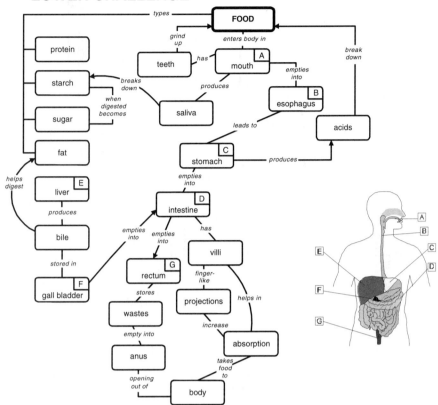

Score: 20 (13 words + 7 letters)

Starting hints: The connectors *grind up* and *fingerlike* provide good starting places.

Notes: This map provides a general look at the digestive process. Because of limited space, some aspects (such as the pancreas, insulin, parts of the small and large intestines, the appendix, etc.) did not make it onto the map. Students seeking extra credit or suggestions for further study could copy this map onto a large paper and then add in the missing elements.

Diagram Key

- **A** mouth
- **B** esophagus
- **C** stomach
- **D** intestine
- **E** liver
- **F** gall bladder
- **G** rectum

HIGHER CHALLENGE

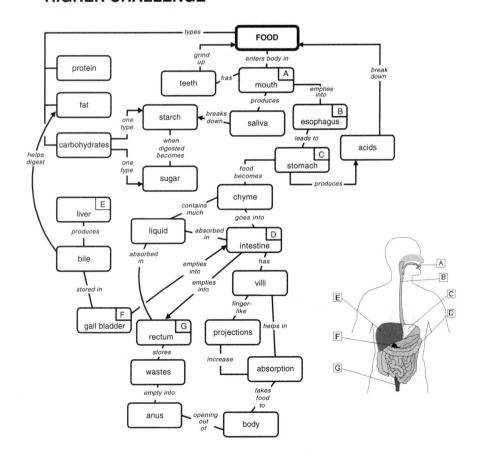

Score: 25 (18 words + 7 letters)

Starting hints: The connectors *grind up* and *fingerlike* provide good starting places.

Notes: This map provides a general look at the digestive process. Because of limited space, some aspects (such as the pancreas, insulin, parts of the small and large intestines, the appendix, etc.) did not make it onto the map. Students seeking extra credit or suggestions for further study could copy this map onto a large paper and then add in the missing elements.

Diagram Key

- **A** mouth
- **B** esophagus
- **C** stomach
- **D** intestine
- **E** liver
- **F** gall bladder
- **G** rectum

Concept Map: Circulatory System

THINKING CONNECTIONS: Life Science Book B — Human Biology

Name _____
Date _____ Period _____

Directions: Select words from the word list and fill in the blank map items. Use each word only once, and use all the words on the list. Then use two different highlighters, colored pencils, or crayons to color in items that are (1) related to oxygen-rich blood, and (2) related to oxygen-poor blood. Show your color scheme in the legend.

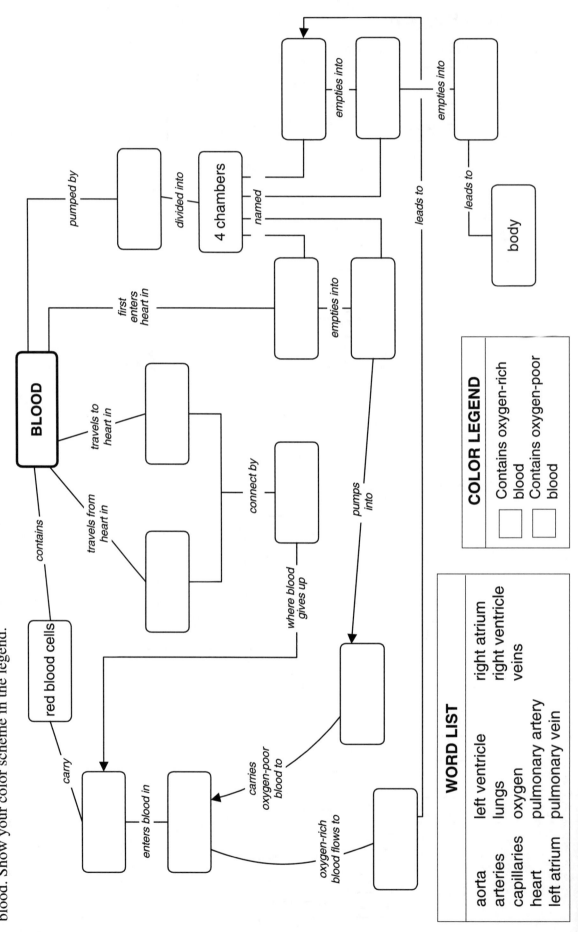

COLOR LEGEND
- ☐ Contains oxygen-rich blood
- ☐ Contains oxygen-poor blood

WORD LIST

aorta	right atrium
arteries	right ventricle
capillaries	veins
heart	
left atrium	
left ventricle	
lungs	
oxygen	
pulmonary artery	
pulmonary vein	

124 © 1998 Critical Thinking Books & Software • www.criticalthinking.com • (800) 458-4849

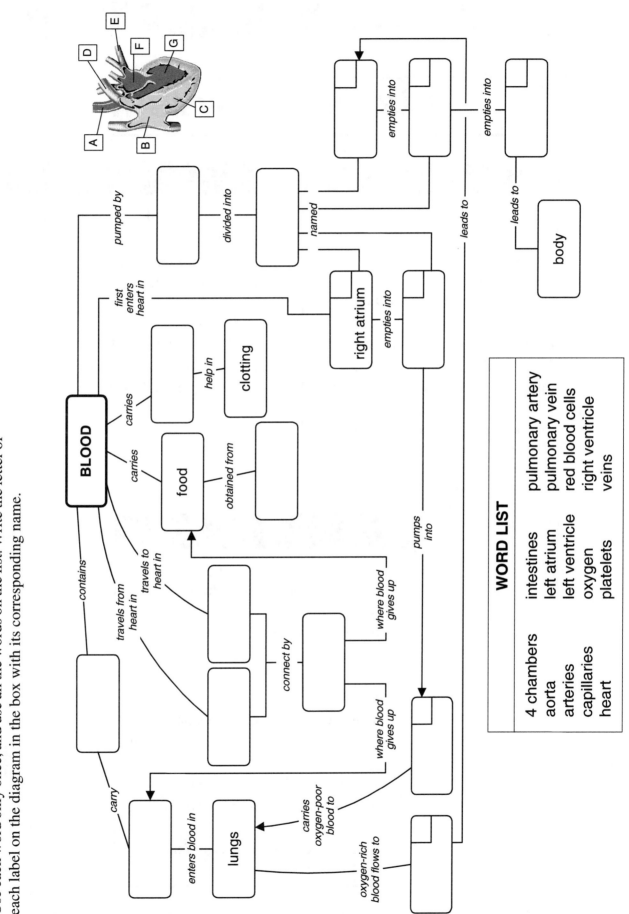

Critical Thinking →

Circulatory System

The Heart

The heart is a hollow muscle that pumps blood throughout the body.

The heart has four chambers:

- left atrium—receives oxygen-rich blood from the lungs through the pulmonary vein
- left ventricle—pumps blood to the aorta
- right atrium—receives blood from the body
- right ventricle—pumps oxygen-poor blood to the lungs through the pulmonary artery

Blood

- contains red blood cells, which carry oxygen
- contains platelets, which help the blood to clot
- travels from the heart in arteries
- travels to the heart in veins
- carries food, picked up from the stomach and intestines

Vocabulary

- ❒ **aorta**—The main artery at the top of the heart that carries blood to the rest of the body.
- ❒ **artery**—A tube that carries blood *away from* the heart.
- ❒ **atrium** (plural: atria)—An upper chamber of the heart.
- ❒ **capillaries**—Tiny tubes that connect arteries to veins; the place where oxygen and food leaves the blood.
- ❒ **pulmonary**—Refers to the lungs.
- ❒ **vein**—A tube that carries blood *to* the heart.
- ❒ **ventricle**—A lower chamber of the heart.

Circulatory System
LOWER CHALLENGE

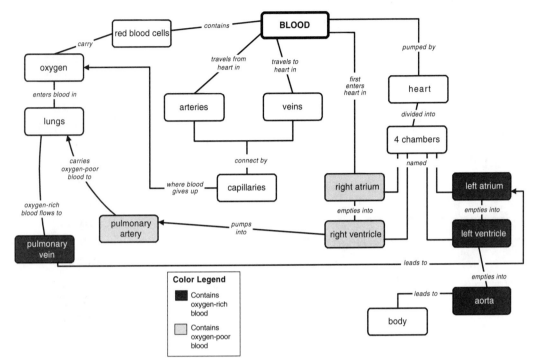

Score: 20 (13 words + 7 colored items)

Starting hints: The seed item *4 chambers* are *named* below it; the top two on each side emptying into the bottom two on each side.

Notes: This map uses a Color Legend to indicate the amount of oxygen in the blood at various stages. If students color in arteries as oxygen-rich, remind them that the pulmonary artery carries oxygen-poor blood. Similarly, if students color in veins as oxygen-poor, remind them that the pulmonary vein carries oxygen-rich blood.

HIGHER CHALLENGE

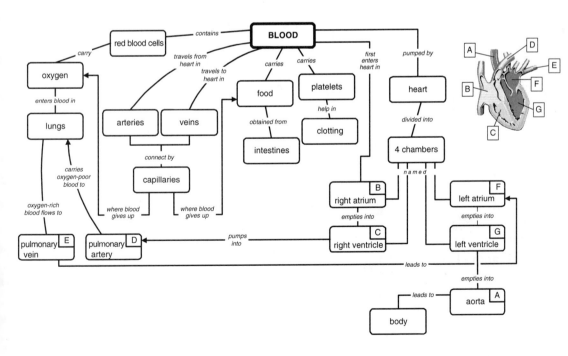

Score: 22 (15 words + 7 letters)

Starting hints: Something (*pulmonary artery*) *carries oxygen-poor blood to* the seed item *lungs*. Something (*platelets*) *help in* the seed item *clotting*.

Notes: This version of the map references a diagram of the heart.

When checking maps, make sure that the letters match the words in the boxes, not necessarily the position on the map.

Students may find it helpful to color the map as above and make their own Color Legend.

Diagram Key

- **A** aorta
- **B** right atrium
- **C** right ventricle
- **D** pulmonary artery
- **E** pulmonary vein
- **F** left atrium
- **G** left ventricle

Concept Map: Respiratory System

Directions: Select words from the word list and fill in the blank map items. Use each word only once, and use all the words on the list.

WORD LIST
- air
- blood
- bronchial tubes
- burning of food
- cartilage
- cells
- heart
- larynx
- lungs
- muscle
- nose
- pharynx
- trachea

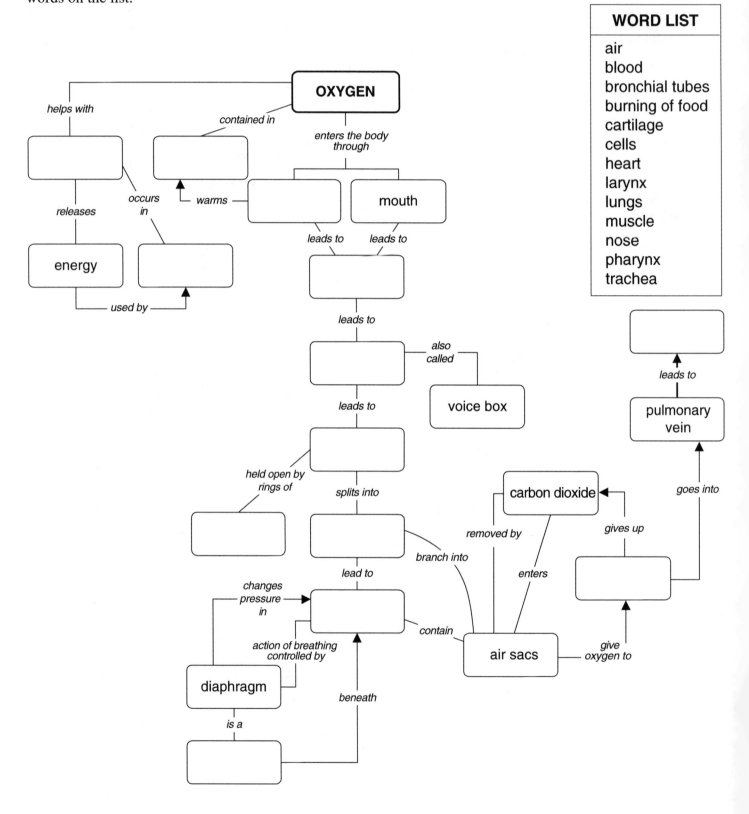

Concept Map: Respiratory System

THINKING CONNECTIONS: Life Science Book B — Human Biology

Name _____
Date _____ Period ____

Directions: Select words from the word list and fill in the blank map items. Use each word only once, and use all the words on the list. Write the letter of each label on the diagram in the box with its corresponding name.

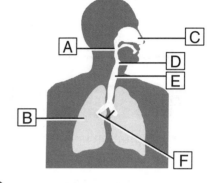

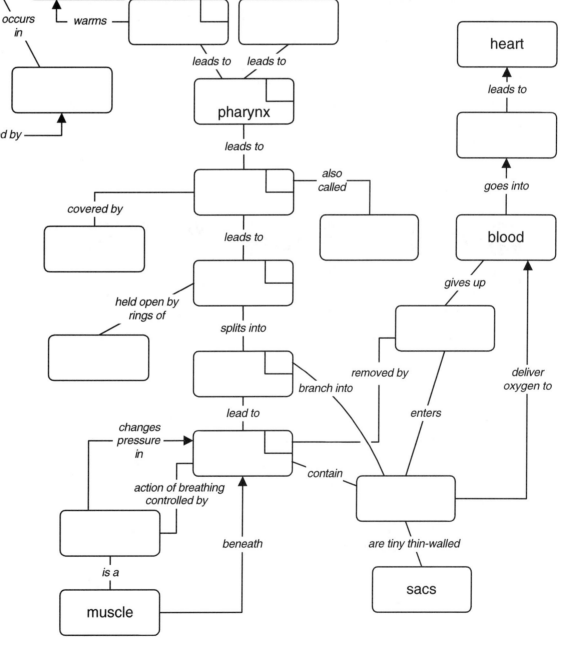

WORD LIST
- air
- alveoli
- bronchi
- burning of food
- carbon dioxide
- cartilage
- cells
- diaphragm
- epiglottis
- larynx
- lungs
- mouth
- nose
- pulmonary vein
- trachea
- voice box

© 1998 Critical Thinking Books & Software • www.criticalthinking.com • (800) 458-4849

THINKING CONNECTIONS: Life Science Book B — Human Biology

Critical Thinking

Respiratory System

Background

- Respiration is the process of burning food and extracting energy.
- Oxygen is necessary for human respiration.
- The lungs allow the transfer of oxygen from the air into the blood.
- The lungs allow the transfer of carbon dioxide, a waste product of food burning, from the blood into the air in the lungs.

Vocabulary

- ❒ **alveoli**—Tiny, thin-walled air sacs in the lungs.
- ❒ **epiglottis**—A flap covering the larynx.
- ❒ **larynx** (LAIR-RINKS)—Part of the air path on which the vocal cords are attached.
- ❒ **pharynx** (FAIR-RINKS)—A passageway for both food and air leading from the mouth.
- ❒ **trachea**—A large tube (sometimes called the windpipe) to the lungs, held open by rings of cartilage.

Breathing

- The muscle in the chest called the diaphragm lowers and expands, creating low pressure in the body.
- Air (containing oxygen) enters the body through the nose and mouth. The nose warms the air on its way in.
- Air passes through the pharynx to the larynx and into the trachea.
- The trachea splits into smaller bronchial tubes.
- The bronchial tubes lead into tiny air sacs in the lungs.
- When the diaphragm contracts, air is pushed out of the lungs and out of the body.

Path of Oxygen and CO_2

- Oxygen in the air in the alveoli moves into the blood.
- Carbon dioxide collects in the blood and moves into the air in the alveoli.
- The oxygen-rich blood around the lungs moves to the heart through the pulmonary vein.
- Oxygen moves to all the living cells of the body where it is used during the process of getting energy from the burning of food.

Respiratory System

LOWER CHALLENGE

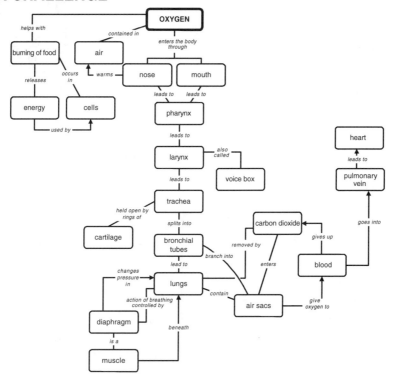

Score: 13 words

Starting hints: Three good starting points would be *air sacs*, *diaphragm*, and *voice box*.

HIGHER CHALLENGE

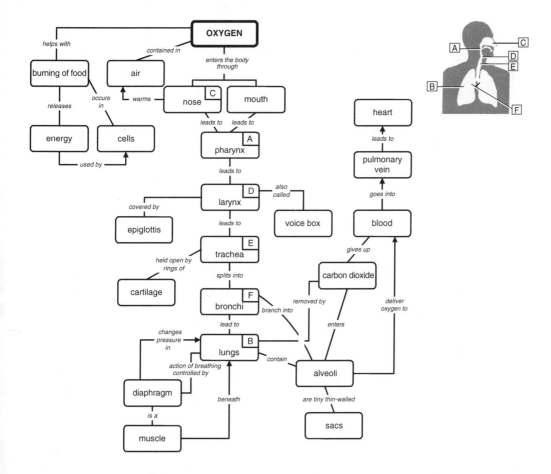

Score: 22 (16 words + 6 letters)

Starting hints: The *alveoli* are tiny *sacs*. The connector *held open by rings of* suggests a ring-lined structure (the *trachea*) and the material of which the rings are made (*cartilage*).

Notes: Some students might reverse *larynx* and *voice box*.

The circulatory and respiratory systems are closely interrelated. Advanced students may find it challenging to take the concept maps of both systems and redraw them so that they are combined. Common terms include *oxygen*, *blood*, *heart*, *lungs*, and *pulmonary vein*.

Diagram Key

- **A** pharynx
- **B** lungs
- **C** nose
- **D** larynx
- **E** trachea
- **F** bronchi

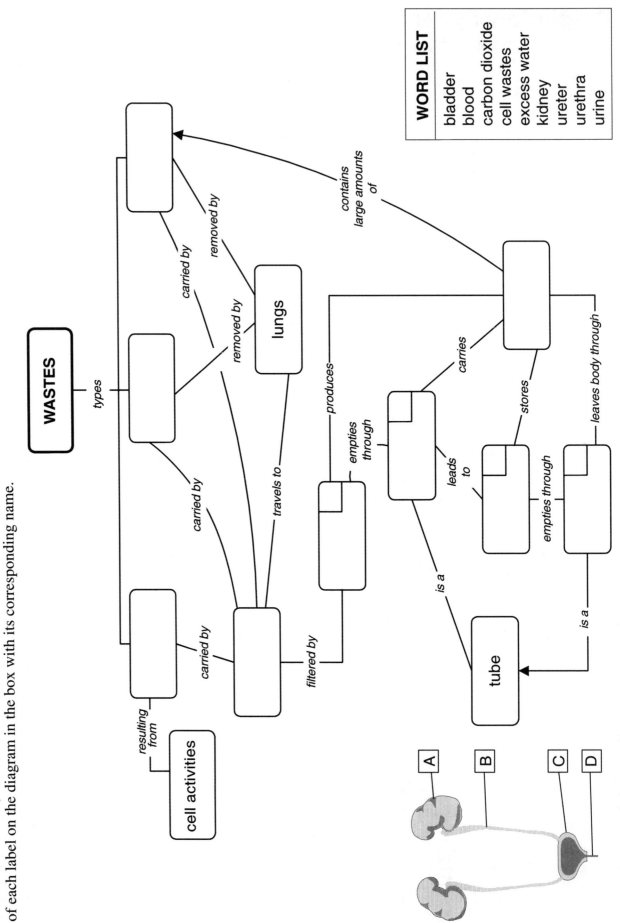

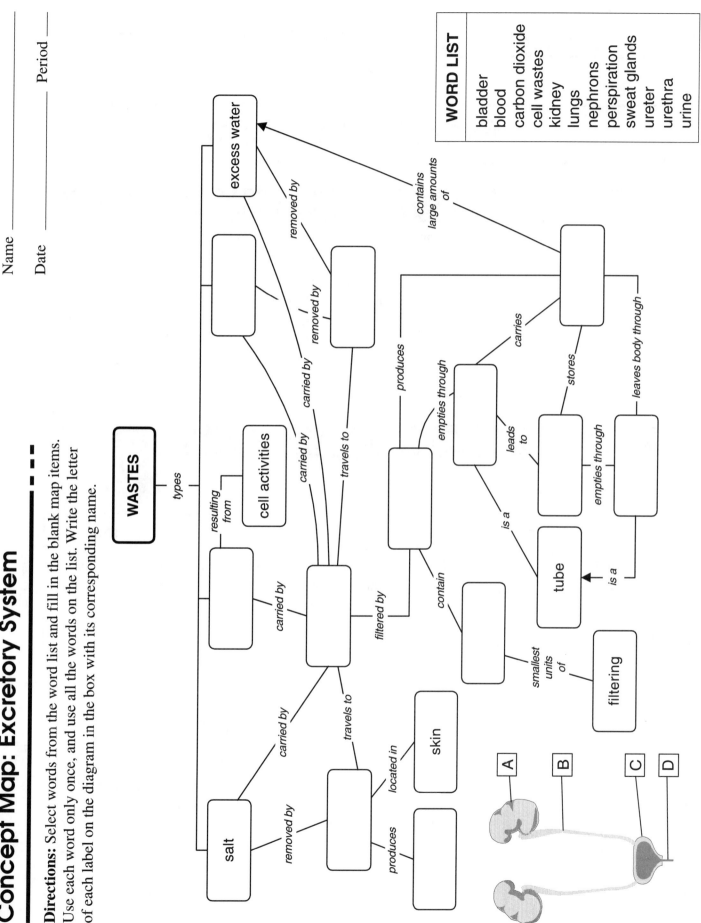

THINKING CONNECTIONS: Life Science Book B — Human Biology

Critical Thinking  Concept File

Excretory System

Background

There are at least four types of wastes in the blood:
- cell wastes produced by cell activity
- excess water in the blood
- salts
- carbon dioxide

Vocabulary

- ❏ **nephron**—The smallest filtering unit of the kidney.
- ❏ **ureter**—The tube that carries urine from the kidney to the bladder.
- ❏ **urethra**—The tube that carries urine from the bladder out of the body.
- ❏ **urine**—A waste liquid containing water, salts, and urea.

Ways that Wastes are Removed

The Skin	The Lungs	The Kidneys
• Skin has sweat glands. • Perspiration helps get rid of 　– salts 　– excess water	• Some wastes enter the lungs and are eliminated with every breath out. 　– carbon dioxide 　– excess water	• The kidneys filter the blood of 　– excess water 　– salts 　– cell wastes • As urine, wastes leave the kidney through the ureter. • Urine is stored in the bladder. • Urine leaves the body through the urethra.

Excretory System

LOWER CHALLENGE

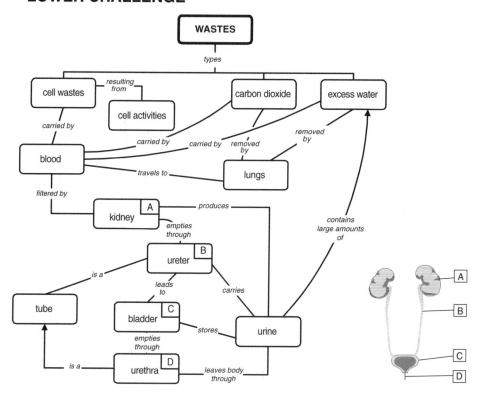

Score: 13 (9 words + 4 letters)

Starting hints: A good starting point is the connector *filtered by*, leading to *kidney*.

Notes: The lower-challenge map does not include the "salt" branch of the higher-challenge.

Also, the lower-challenge map shows which blocks will receive diagram letters, whereas the higher-challenge map does not.

Diagram Key
- A kidney
- B ureter
- C bladder
- D urethra

HIGHER CHALLENGE

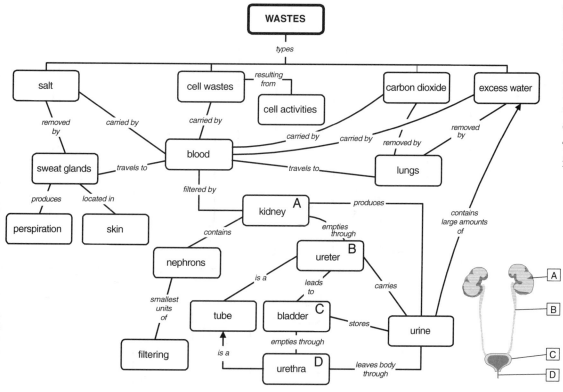

Score: 16 (12 words + 4 letters)

Starting hints: A good starting point on these maps is the connector *filtered by*, leading to *kidney*.

Notes: The higher-challenge map does not indicate which blocks will receive diagram letters.

The higher-challenge map includes a "salt" branch.

Diagram Key
- A kidney
- B ureter
- C bladder
- D urethra

Concept Map: Nervous System

Directions: Select words from the word list to fill in the blank map items. Use each word only once, and use all the words on the list.

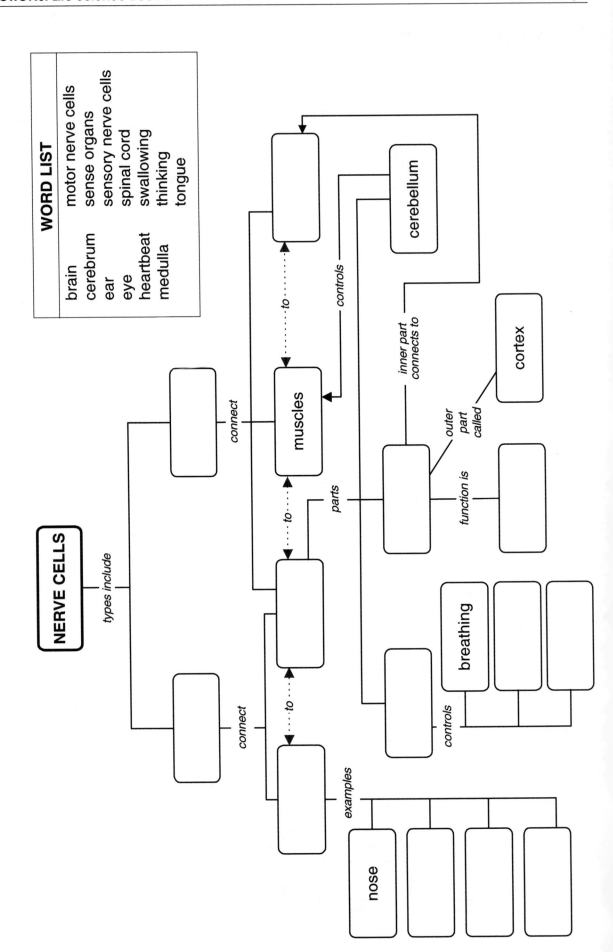

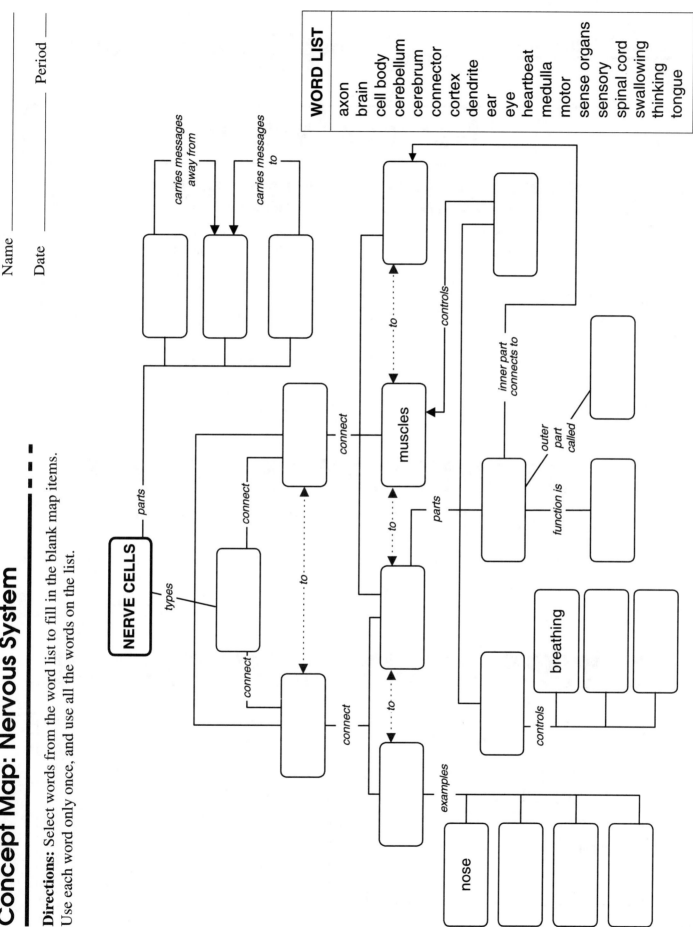

THINKING CONNECTIONS: Life Science Book B — **Human Biology**

Critical Thinking →

Nervous System

Characteristics

The nervous system includes the brain, the spinal cord, and a network of nerve cells.

Parts of a nerve cell are
- axon
- cell body
- dendrite

Vocabulary

- ❒ **axon**—The part of a nerve cell that carries impulses *away from* the cell.
- ❒ **connector nerve cells**—Connect sensory neve cells to motor nerve cells.
- ❒ **cortex**—The outer part of the cerebrum.
- ❒ **dendrite**—The part of a nerve cell that carries impulses *into* the cell.

Parts of the Brain

- medulla—controls involuntary responses such as the heartbeat, swallowing, and breathing
- cerebrum—controls mental processes such as thinking
- cerebellum—controls muscular actions

Major Types of Nerve Cells

Sensory Nerve Cells
- connect sense organs to the brain
- sense organs include
  – nose
  – ear
  – eye
  – tongue

Motor Nerve Cells
- function in voluntary or involuntary movement of muscles
- connect muscles to the
  – brain
  – spinal cord

Nervous System

LOWER CHALLENGE

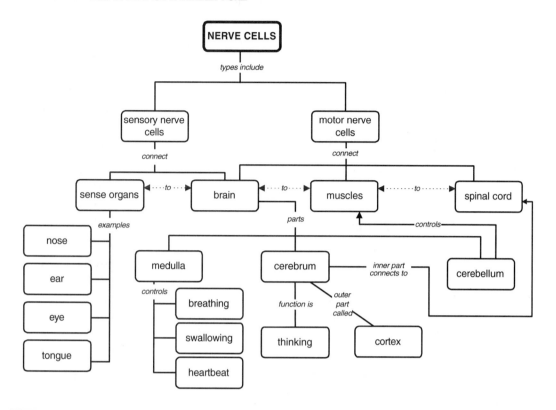

Score: 13 words

Starting hints: The items *sensory nerve cells* and *motor nerve cells* are obvious types of nerve cells.

The seed item *nose* should indicate a list of sense organs, and the seed item *breathing* should indicate a list of involuntary processes. The only item with an outer part on the list is the *cerebrum*.

Notes: Sense organs and processes controlled by the medulla can be listed in any order.

HIGHER CHALLENGE

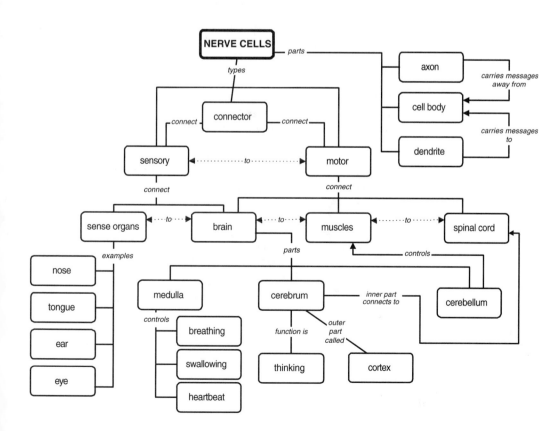

Score: 19 words

Starting hints: The seed item *nose* should indicate a list of sense organs, and the seed item *breathing* should indicate a list of involuntary processes. The only item with an *outer part* on the list is the *cerebrum*.

The connectors to *cell body* give ample clues as to the items around it.

Notes: Sense organs and processes controlled by the medulla can be listed in any order.

THINKING CONNECTIONS: Life Science Book B Human Biology

Concept Map: Human Reproduction

Name _____

Date _____ Period ____

Directions: Select words from the word list and fill in the blank map items. Use each word only once, and use all the words on the list. Then use two different highlighters, colored pencils, or crayons to color in items that are (1) parts of the male reproductive system and (2) parts of the female reproductive system. Show your color scheme in the legend.

WORD LIST
- amniotic sac
- birth canal
- embryo
- fetus
- ovary
- oviduct
- ovulation
- semen
- sperm
- testis
- uterus

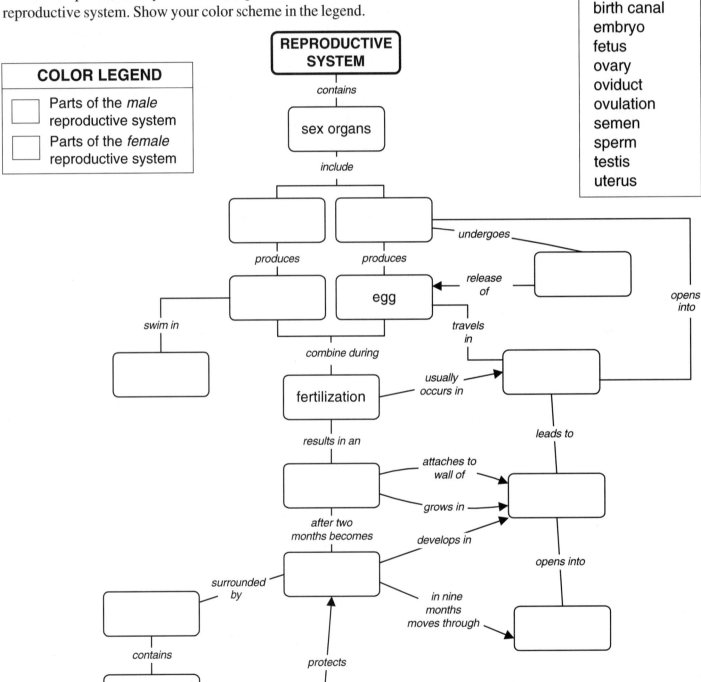

Concept Map: Human Reproduction

Directions: Select words from the word list and fill in the blank map items. Use each word only once, and use all the words on the list. Write the letter of each label on the diagram in the box with its corresponding name.

WORD LIST

amniotic fluid, ovary, amniotic sac, oviduct, cushioning, semen, embryo, sex organs, female system, sperm, fertilization, testis, fetus, uterus, male system, vagina

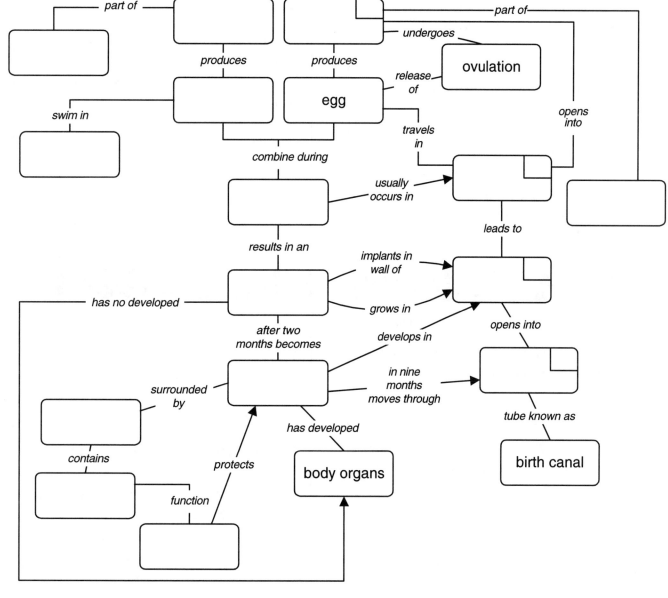

THINKING CONNECTIONS: Life Science Book B — Human Biology

Critical Thinking

Human Reproduction

Sequence

- Sperm and egg usually join in the oviduct and form a zygote.
- The zygote travels to the uterus.
- The zygote divides several times, producing more cells and becoming an embryo.
- The embryo attaches to the wall of the uterus.
- After two months of development in the uterus, the embryo grows and becomes a fetus.
- The fetus is enclosed by an amniotic sac filled with amniotic fluid, which cushions and protects the fetus.
- After a total of nine months, the uterus pushes the fetus to the outside through the birth canal.

Vocabulary

- ❏ **fetus**—the stage of development after two months' growth, characterized by the appearance of body organs.
- ❏ **fertilization**—The joining of sperm and egg.
- ❏ **flagellum** (plural: flagella)—A whiplike structure used during movement and made of fibers of protein.
- ❏ **embryo**—The early stages of development after fertilization, characterized by the lack of body organs.

Male

- Sperm cells are the male gametes.
- Sperms have a flagella and carry one half of the hereditary traits.
- Sperm cells are produced in the testes (singular: testis).
- Sperm cells are combined with semen, a fluid in which sperm swim.
- Fertilization usually (but not always) occurs in the oviducts.

Female

- The egg is the female gamete.
- Eggs carry one half of the hereditary traits.
- Egg cells are produced in the ovaries.
- During ovulation, an egg is released from the ovary.
- The egg travels down the oviduct to the uterus.
- A fertilized egg will implant itself in the wall of the uterus.